The Influence of Different Sources of Visual Information on Joint Action Performance

Dissertation

zur Erlangung des Grades eines
Doktors der Naturwissenschaften

der Mathematisch-Naturwissenschaftlichen Fakultät
und der Medizinischen Fakultät
der Eberhard-Karls-Universität Tübingen

vorgelegt von:

Stephan Streuber

aus Lutherstadt Wittenberg, Deutschland

Juli - 2012

Bibliografische Information der Deutschen Nationalbibliothek

Die Deutsche Nationalbibliothek verzeichnet diese Publikation in der
Deutschen Nationalbibliografie; detaillierte bibliografische Daten sind
im Internet über http://dnb.d-nb.de abrufbar.

ISBN 978-3-8325-3373-1

Logos Verlag Berlin GmbH
Comeniushof, Gubener Str. 47,
10243 Berlin
Tel.: +49 (0)30 42 85 10 90
Fax: +49 (0)30 42 85 10 92
INTERNET: http://www.logos-verlag.de

Tag der mündlichen Prüfung: 15.02.2013

Dekan der Math.-Nat. Fakultät: Prof. Dr. W. Rosenstiel
Dekan der Medizinischen Fakultät: Prof. Dr. I. B. Autenrieth

1. Berichterstatter: Prof. Dr. Heinrich H. Bülthoff
2. Berichterstatter: Prof. Dr. Günther Knoblich

Prüfungskommission: Prof. Dr. Heinrich H. Bülthoff
                             Prof. Dr. Günther Knoblich
                             Prof. Dr. Uwe Kloos
                             Dr. Hong Yu Wong

Ich erkläre, dass ich die zur Promotion eingereichte Arbeit mit dem Titel: *„The Influence of Different Sources of Visual Information on Joint Action Performance"* selbstständig verfasst, nur die angegebenen Quellen und Hilfsmittel benutzt und wörtlich oder inhaltlich übernommene Stellen als solche gekennzeichnet habe. Ich versichere an Eides statt, dass diese Angaben wahr sind und dass ich nichts verschwiegen habe. Mir ist bekannt, dass die falsche Abgabe einer Versicherung an Eides statt mit Freiheitsstrafe bis zu drei Jahren oder mit Geldstrafe bestraft wird.

Stephan Streuber

Acknowledgements

First and foremost I would like to thank Stephan de la Rosa for his support and supervision. It was always constructive, inspiring and most enjoyable to work with Stephan over the last couple of years. I cannot imagine a better supervisor.

Further, I would like to thank Heinrich H. Bülthoff who was always supportive of my scientific work at the Max Planck Institute for Biological Cybernetics and who gave me the opportunity to use all the research facilities and to present my work at international conferences. I also would like to thank Betty Mohler, Günther Knoblich, Natalie Sebanz, Joachim Tesch, Trevor Dodds, Tobias Meilinger and all my colleagues form the MPI for their support of my scientific work. Without them my research would have not been possible.

A special thanks to Florian Soyka, Betty Mohler, Eva Hanau, Trevor Dodds, Shih-pi Ku, Zoltan Borlan, Benjamin Kummer, Regine Armann, Julia Denz, Ines Wolff, Paolo Robuffo Giordano, Dong-Seon Chang, and Gabriela Wacker for making my time at the MPI and in Tübingen very enjoyable.

Most of all, I feel grateful to my parents, who have always supported me throughout my life.

Abstract

Humans are social beings and they often act jointly together with other humans (joint actions) rather than alone. Successful joint action requires the understanding and the coordination of one's own actions with another person's actions. The current Ph.D. thesis attempts to advance our knowledge about joint action coordination by extending existing research in two novel and important ways. First, prominent theories of joint action agree on visual information being critical for successful joint action coordination but are vague about the exact source of visual information being used during a joint action. However, in a real life interaction several sources of visual information exist which inform an interaction partner about the ongoing course of the interaction (e.g. visual information about objects, tools, other persons). Knowing which sources of visual information are used, however, is important for a more detailed characterization of the functioning of action coordination in joint actions. Second, previous studies investigating the role of visual information in social settings often constrain the experimental tasks to artificial laboratory settings. To examine joint action mechanisms under realistic conditions I devised experimental tasks that allowed a close-to-natural joint action. As a result the perceptual and motor components of the experimental tasks were less constrained allowing for a more natural interaction compared to previous studies. The current Ph.D. research examines the importance of different sources of visual information on joint action coordination under realistic settings. In three studies I examined the influence of different sources of visual information (Study 1), the functional role of different sources of visual information (Study 2), and the effect of social context on the use of visual information (Study 3) in a table tennis game. The results of these studies revealed that (1) visual anticipation of the interaction partner and the interaction object is critical in natural joint actions, (2) different sources of visual information are critical at different temporal phases during the joint action, and (3) the social context modulates the importance of different sources of visual information. In sum, this work provides important and new empirical evidence about the importance of different sources of visual information in close-to-natural joint actions.

Table of Contents

Introduction

Joint Action Coordination

An integral part of human life is the interaction with other human beings (social interaction). A key feature of many social interactions is that the actions of two persons must be coordinated in a non-verbal fashion in space and time, such as when handing over an object to another person, fixing a bike together, dancing a tango or playing a table tennis game. We will refer to this class of social interactions as joint actions (Sebanz, Bekkering, & Knoblich, 2006).

A fundamental difference between acting alone and acting jointly is that actions must not only be coordinated with the physical environment but also with the actions of others (Sebanz, et al., 2006). For instance, when carrying a wounded soccer player from the field together with a stretcher both helpers must coordinate what kind of action needs to be performed (e.g. lifting the stretcher), the precise time when this action must be performed (e.g. the time the stretcher must be lift up) and where this action must be performed in space (e.g. walk to a specific location) in a way such that the wounded player does not fall from the stretcher. The "What", "When", and "Where" aspect of joint action must be coordinated in real-time and on the basis of perceptual information about the physical and social environment (Sebanz & Knoblich, 2009). Despite the required amount of interpersonal coordination, humans often coordinate their actions with ease and without discussions, diagrams, or intense thinking. How do humans achieve such smooth and efficient joint action coordination, for example when they play a table tennis game?

Online motor control and social cognition are two research fields that provide partial answers to this question. Research on online motor control is primarily interested in the question how the central nervous system continuously converts sensory inputs into motor outputs when interacting within a dynamically changing environment, such as when catching a fly ball (see e.g., reviews by Turvey, 1990; Warren, 2006). On the other hand theories on social cognition are interested in the question how humans grasp the minds of others and how this ability affects individual's behavior within a

3

social context (see e.g., reviews by U. Frith & Frith, 2010; Ochsner & Lieberman, 2001; Schilbach et al., 2012). Importantly, research on online motor control is often interested in the mechanisms behind object interaction while social cognition research mainly focuses on the mechanisms behind the interaction of two people regardless of the presence of an object. In everyday life, however, humans often interact with physical objects and with other humans simultaneously. For instance when handing over a cup of coffee to another person, both interaction partners need to plan and execute their motor movements with respect to physical properties (e.g. the position of the cup) but also with respect to social properties (e.g. what they think the other person intends to do with the cup). Hence, understanding object interactions within social settings requires the understanding of the processes as described by theories on online motor control and theories on social cognition and possibly their interactions.

A particular type of joint action that seems to be well suited for the investigation of the contribution of online motor control and social cognition on behavior are rapid ball games such as table tennis, badminton or baseball (Van der Kamp, Rivas, Van Doorn, & Savelsbergh, 2008). Successful performance in those types of sports games requires fine spatio-temporal action coordination in order to place the ball to an intended location. Sportsmen exhibit a high degree of temporal and spatial accuracy in those joint actions. For instance, table tennis players hit approaching table tennis balls within a temporal window of 7 ms when they repeatedly respond to the same ball trajectory (Bootsma & Vanwieringen, 1990). It has been argued that such accurate interceptions requires visual anticipation (Van der Kamp, et al., 2008), which is humans' ability to make accurate predictions of a physical event from incomplete, temporally preceding sources of visual information (Poulton, 1957). Until now, however, it is not entirely clear, what sources of visual information are actually anticipated in order to make those predictions. This piece of evidence can be used to begin to dissociate the contribution of online motor control and social cognition within joint actions.

In the next two paragraphs I will outline the possible contributions of online motor control mechanisms and social cognitive mechanisms on joint action.

Online Motor Control

Interacting with an object often requires humans to coordinate their body movements in real-time with the physical world based on the sensory information about physical objects (Bootsma & Vanwieringen, 1990; Grierson, Gonzalez, & Elliott, 2009; McBeath, Shaffer, & Kaiser, 1995; McLeod & Dienes, 1993a; Sarlegna & Blouin, 2010). For example, baseball players adjust their catching behavior to disturbances of the baseball's flying trajectory induced while they are catching the ball (Fink, Foo, & Warren, 2009). Other evidence for real-time action coordination by means of direct visual feedback comes from the observation that humans adjust their arm movements to disturbances of the object position induced while they are grasping an object (Reichenbach, Thielscher, Peer, Bülthoff, & Bresciani, 2009). Control laws describe how the central nervous system continuously converts sensory inputs into motor outputs (see reviews by Turvey, 1990; Warren, 2006). These studies suggest a close link between visual information and the online control of motor movements when humans interact with physical objects. Within this framework a coupling of body movements and visual information in an online manner provides a fast and robust strategy to deal with perturbations and noise. How can online motor control aid joint action coordination? Several joint action tasks have properties that make online control appear as a candidate mechanism for action coordination. Specifically joint tasks that rely on visual information in a time critical manner could benefit from online control mechanisms. For instance, many sports games require players to catch or intercept flying balls with a racket in a time critical manner. Visual information about the flying ball is likely to provide the critical source of visual information in order to control movements when catching or intercepting flying balls. For instance, in the case of parabolic ball flight, the only thing a player needs to do in order to intercept the ball is to null the vertical optical acceleration of the balls projection on the retina (Chapman, 1968; Fink, et al., 2009; McLeod & Dienes, 1996; McLeod, Reed, & Dienes, 2006; McLeod, Reed, Gilson, & Glennerster, 2008). Several suggestions exists how this close coupling between action and perception can be achieved: by means of physical knowledge about the world or as a direct perception-action coupling that is based on perceptual invariants. A direct coupling between visual ball information and motor movements has been observed in many joint actions such as table tennis (Bootsma & Vanwieringen, 1990), baseball

5

(Fink, et al., 2009) and soccer (McLeod, et al., 2008). Online motor control might aid time-critical joint actions by allowing rapid real-time control of behavior on the basis of instantaneous visual information (e.g. when catching a baseball or when blocking a karate kick). Hence, theories on online motor control suggest that visual information is important immediately. On the other hand, visual information that is older than a few hundred milliseconds should be of little use.

Social Cognition

An integral part of human life is the interaction with other human beings. A major concern of social cognition research is how human behavior is affected by the ability to grasp the internal states of others (e.g. intentions, beliefs, plans, emotions) during a social interaction. The importance of the other person's (unknown) internal states for action coordination rests in their ability to have a direct effect on action coordination. An example for a situation in which an incorrect interpretation of internal states affects action coordination is when two friends would like to greet each other but one is intending to hug while the other one is intending to shake hands. Hence estimating and predicting the other person's internal states is important for successful joint action (Adolphs, 1999, 2003; Sebanz, et al., 2006). One challenge that humans face when coordinating actions with those of others is that the other person's internal states (e.g. intentions, action plans) are unknown. Therefore a key question of social cognition research is how do humans determine the internal states of others and how does this knowledge affect human behavior when interacting with others? One obvious way of knowing the internal states of others is by communicating them explicitly using language. Verbal communication is a powerful tool for coordinating joint actions (Clark, 1996). However, in everyday life verbal communication is often too slow for time critical action coordination or simply not possible. For instance, in a soccer game it defeats the purpose of surprise and it is too time consuming to coordinate each individual players action verbally in order to jointly attack the opponent's goal.

Another way of inferring internal states of others is by observing the other person's behavior. The general ability to understand and infer the mental states of others has been referred to as theory of mind (Flavell, 2004; Premack & Woodruff, 1978). Different theories have been proposed about how humans attribute mental states to others. One can broadly distinguish between so-called 'theory theories' and 'simulation theories'. The 'theory theory' on action understanding assumes that humans generalize their own experience and develop a naïve theory - similar to an ordinary scientific explanation – in order to explain actions of third persons (Churchland, 1988; Fodor, 1987; C. D. Frith & Frith, 2006; U. Frith & Frith, 2010; Schilbach, et al., 2012; Sellars, 1956). Hence, the 'theory theory' on action understanding proposes that knowledge about others' mental states is inferential, reflective, and explicate. Instead, the 'simulation theory' on action understanding, also known as 'motor simulation theory' states that actions of others are understood by using one's own motor system to adopt the other's perspective (Gallese & Goldman, 1998; Rizzolatti & Sinigaglia, 2010; Schilbach, et al., 2012; Sellars, 1956). Two distinct large-scale neuroanatomical networks have been associated with humans ability to grasp other's minds by observing their behavior. The 'mirror neuron system' is associated with the 'simulation theory' of action understanding and the 'mentalizing network' has been associated with the 'theory theory' of action understanding. So far it has remained unclear whether and to what extent the neural activity observed in these networks serve complementary or mutually exclusive roles in action understanding (Schilbach, et al., 2012). Importantly (and critical for the present thesis) both theories on action understanding assume that the observed behavior of another person is one possible input to the 'simulation' and 'theory theory' mechanism. Hence, they suggest that humans can infer the mental states of others by observing their behaviour.

Evidence for human's ability to understand the intentions underlying other's actions by means of observing their behavior comes from behavioral studies. For example, humans are able to identify the intentions underlying observed body movements from point light stimuli (Barrett, Todd, Miller, & Blythe, 2005; Runeson & Frykholm, 1983). Point lights are devoid of figural cues but preserve the essential movement kinematics of an action (Johansson, 1973). In previous research, point-light stimuli were presented on video displays in order to demonstrate observer's ability to detect the kind of actions performed (Dittrich, 1993; Vanrie & Verfaillie, 2004), to recognize the actor's expectations (Runeson & Frykholm, 1983), and to understand the actor's intentions (Grezes, Frith, & Passingham, 2004a). Inferring the other person's intentions is important for joint actions because observers can predict what the other person is going to do next. For example, goal keepers can better predict the fate of a penalty kick when observing the body of the penalty kicker prior to ball contact (Savelsbergh, Williams, Van der Kamp, & Ward, 2002). Also, basketball players can better predict the fate of a basketball shot when observing the body of the shooter (Aglioti, Cesari, Romani, & Urgesi, 2008) before the ball is released from the hand. In sum social cognition might aid joint actions by allowing interaction partners to understand and predict each other's behavior on the basis of visual information. Therefore, theories on social cognition suggest that visual information is important at an early point of time during the interaction (e.g. before an opponent hits the ball in a table tennis game).

The Aim and Structure of the Thesis

The aim of this thesis was to investigate the contribution of visual information on joint action under close-to-natural conditions. The previous sections discussed the possible contributions of online motor control and social cognitive processes to joint action coordination. Specifically, theories on online motor control suggest that visual information should aid time-critical joint actions by allowing rapid real-time control of behavior on basis of instantaneous visual information. On the other hand theories on social cognition stress the importance of visual information about the interaction partner's actions for understanding and predicting the other's internal states. Hence, theories on online motor control and theories on social cognition suggest that visual information should guide behavior in a joint action task. However, both theories differ with respect to the importance of specific sources of visual information and the time, when these sources of visual information become critical.

Previous research often investigated both mechanisms (online motor control and social cognition) in isolation. For instance, research on online motor control investigated human behavior in interceptive tasks (e.g. catching or hitting a flight ball) irrespective of a social context (Bootsma & Vanwieringen, 1990; Fink, et al., 2009; McLeod, et al., 2008). On the other hand research on social interaction often investigated the effect of social stimuli on behavior and brain activity irrespective of physical objects (Dittrich, 1993; Vanrie & Verfaillie, 2004). However, many real life tasks require that objects are coordinated between two agents. It seems likely that these tasks require an interaction between motor control and social cognition.

8

Moreover, to gain better understanding about how mechanisms as described in social cognition and online motor control research interact over time it is important to examine joint action under close-to-natural conditions. Hence, this thesis aims to investigate joint action under close-to-natural conditions. Previous research was often based on the assumption that the processes underlying joint action (e.g. visual anticipation) can be understood using artificial laboratory tasks that might be very different from natural joint actions. For instance, researchers investigated visual anticipation in social settings using psychophysical methods in which participants were required to press buttons in response to social stimuli presented on a computer screen (Abernethy, 1988, 1990; Abernethy & Russell, 1987; Abernethy & Zawi, 2007; Abernethy, Zawi, & Jackson, 2008; Aglioti, et al., 2008; Farrow & Abernethy, 2003; Houlston & Lowes, 1993; Huys et al., 2009; Isaacs & Finch, 1983; McMorris & Colenso, 1996; Penrose & Roach, 1995; Renshaw & Fairweather, 2000; Salmela & Fiorito, 1979; Urgesi, Savonitto, Fabbro, & Aglioti, 2011; Weissensteiner, Abernethy, Farrow, & Muller, 2008). This experimental paradigm provides a high degree of control over the factors in question but is based on the assumption that perception is a process of assigning meaning to visual information via inferential processes (Davids, 2002; Williams, Davids, & Williams, 1999). Hence, perception was investigated without regard to its function (e.g. action). However, it has been argued that visual perception is not the passive intake of visual information but the detection of specific optic variables or invariants that specify a forthcoming action (perception for action) (Wolpert, Ghahramani, & Flanagan, 2001). This suggests that visual perception must be studied with respect to its function for action. This view emphasizes the importance of a functional coupling between action and perception. It suggests that restricting action (e.g. by forcing participants to press a button) should alter perception and vice versa. Therefore, constraining the joint actions within the experimental settings should reduce ecological validity of the studies by not preserving the natural coupling between perception and action (Van der Kamp, et al., 2008). In order to address this important aspect I aimed to investigate the role of visual information under close-to-natural conditions, preserving the natural linkage between action and perception.

In order to systematically investigate the contribution of different sources of visual information to join action behavior (see Study 1 & Study 2) under more natural

9

conditions, I developed a novel experimental environment that allowed a temporal manipulation of the visibility of different sources of visual information while a participant engages in a natural joint action task. In this virtual environment (VE) participants played table tennis with a virtual table tennis player, while visual information was selectively manipulated during play. Using a VE allowed me to immerse participants in a computer-generated three-dimensional world. In this world, participants saw a stereoscopic image of a table tennis situation from an egocentric perspective through a head mounted display. An optical tracking system was used to track the participant's racket and his/her head motion in real time. The virtual table tennis opponent was a computer generated avatar which executed table tennis strokes towards the participant. The participant's task was to hit the virtual ball with a table tennis racket. In the VE the natural coupling between action and perception was persistent allowing participants to perform table tennis strokes similar as in the real world. Hence, the VE allowed for more natural interaction than previous studies using psychophysical occlusion techniques. For instance, previous studies using psychophysical occlusion paradigms assessed performance in terms of button presses in response to video clips shown on a computer screen (e.g. Abernethy, 1988, 1990; Abernethy & Russell, 1987; Abernethy & Zawi, 2007; Abernethy, et al., 2008; Aglioti, et al., 2008; Farrow & Abernethy, 2003; Houlston & Lowes, 1993; Huys, et al., 2009; Isaacs & Finch, 1983; McMorris & Colenso, 1996; Penrose & Roach, 1995; Renshaw & Fairweather, 2000; Salmela & Fiorito, 1979; Urgesi, et al., 2011; Weissensteiner, et al., 2008). Constraining the response action to a button press makes the response the result of a cognitive explicit task rather than an online motor response (implicit task) that is closely coupled to the other person's actions. Further, button presses forced participants to judge the fate of an observed action only after anticipation occurs. Assessing the 3D-movement kinematics of the table tennis strokes allowed me to access participant's behavior over the entire time course of the interaction, also in the moment when visual information should effect participant's behavior (e.g. when the opponent hit the ball). In sum, the VE allowed for the same degree of control as in previous psychophysical occlusion paradigms while at the same time allowing for a natural linkage between action and perception. Thus, I anticipated the results to be more informative and ecologically valid than the results from the studies using psychophysical occlusion techniques.

10

In Study 1 the VE was employed to systematically investigate the contribution of different sources of visual information on joint action behavior when engaging in table tennis play. Specifically, I tested the hypothesis that different sources of visual information (e.g. about the ball, the paddle and the virtual player) guide participant's motor movements in a table tennis game (e.g. visual information about the ball, the paddle and the body of the opponent). This hypothesis was derived from theories on online motor control and theories on social cognition. For instance, theories on online motor control suggest that visual information about the ball is most critical because it allows the rapid real-time control of behavior on basis of instantaneous visual information. On the other hand theories on social cognition often stress the importance of visual information about the interaction partner's body for understanding and predicting the other's internal states. Hence, theories on online motor control and theories on social cognition suggest that different sources of visual information should guide behavior in a joint action task. In order to test to which extent this hypothesis holds true I asked participants to hit table tennis balls in response to a virtual player while visual information was selectively manipulated. I tested the effect of the visibility of the ball, the paddle and the body of the virtual player on task performance and movement kinematics. Task performance was measured as the minimum distance between the center of the participant's paddle and the center of the ball (radial error). Movement kinematics was measured as variability in the paddle speed of repeatedly executed table tennis strokes (stroke speed variability). I found that radial error was reduced when the ball was visible compared to invisible. However, seeing the body and/or the racket of the virtual player only reduced radial error when the ball was invisible. There was no influence of seeing the ball on stroke speed variability. However, I found that stroke speed variability was reduced when either the body or the paddle of the virtual player was visible. Importantly, the differences in stroke speed variability were largest in the moment when the virtual player hit the ball. This suggests that seeing the virtual player's body or paddle might be important for preparing the stroke response. These results demonstrate for the first time that the control of arm movements is not only guided by visual ball information but also by visual information about an interaction partner's paddle and body. However, Study 1 did not allow an assessment of the contribution of a specific source of visual information (e.g. about the

11

ball) for a specific functional role within a joint action as described by theories on online motor control and theories on social cognition. Hence, in order to dissociate the contribution of specific sources of visual information for online motor control and social cognition I designed Study 2.

Study 2 used the VE in order to further dissociate the contribution of specific sources of visual information for online motor control processes and social cognitive processes in joint action. Specifically, I tested the hypothesis that different sources of visual information have their maximum importance for joint action at different temporal phases during the interaction. This hypothesis was derived from theories on online motor control and theories on social cognition. For instance, theories on social cognition suggest that visual information about the other player is used to predict the action of the virtual player before he returns the ball. Consequently, visual information about the other player should be important at the time when the other player hits the ball or even before (i.e. early within the interaction). On the other hand, theories on online motor control suggest that visual information about the ball is used to guide paddle-ball coordination. As a result the ball information might be important shortly before the participant hits the ball. I tested this hypothesis by temporally occluding different sources of visual information during table tennis play. This allowed for an assessment of the temporal phase in which each source of visual information was most critical for participant's performance (in terms of radial error and directional accuracy). The results revealed that visual information about the ball was most critical at a later temporal phase of the interaction, as suggested by theories on online motor control. On the other hand seeing the virtual player's body and paddle was most critical before the virtual player returned the ball as suggested by theories on social cognition. Hence, the results are in line with the initial hypothesis that different sources of visual information are critical at different temporal stages of the interaction.

In sum, using a novel experimental paradigm (the VE) allowed me to investigate visual anticipation under better ecological validity maintaining the natural linkage between action and perception. The results of Study 1 and Study 2 are in line with the idea that humans anticipated different sources of visual information in order to derive short-term predictions based on visual ball information and in order to derive long-term predictions based on visual information about the interaction partner. The results of Study 1 and

Study 2 were in line with the predictions of theories on online motor control and theories on social cognition. Here I show evidence of how these two sources of visual information interplay, under more ecologically valid conditions. Thus, these findings provide new insights into the contribution of visual information on joint action behavior.

On the other hand, Study 1 and Study 2 are missing reciprocal interactions, which typically occur in natural joint actions. That is, in natural joint actions both interaction partners reciprocally influence each other's actions (closed loop). Hence, natural joint actions might not only require visual information in order to derive predictions but also in order to plan own actions with respect to how these actions affect the opponent's play. For instance, hitting a ball close to the net will force the opponent player to step closer to the table. Similar, outperforming an opponent usually requires a subsequent number of specific strokes that will stepwise reduce the action opportunities at the disposal of the opponent player. Visual information might be critical for this task. For instance, visual information might be critical in order to decide where to strike a ball so that the opponent cannot reach the ball. Such an effect would be "invisible" using "isolation paradigms" (Becchio, Sartori, & Castiello, 2010) which do not take into account the reciprocity of joint action. The need for leaving the "experimental quarantine" of "isolation paradigms" in order to allow for reciprocal joint action in a "closed loop" has been recently recognized (Becchio, et al., 2010; U. Frith & Frith, 2010; Schilbach, et al., 2012; Sebanz, et al., 2006). These authors argue against the assumption that perception, action, and higher-level cognitive processes can be understood by investigating individual minds in isolation (Sebanz, et al., 2006). Consequently, these authors point to the necessity for the investigation of reciprocal interactions embedded in a social context (Becchio, et al., 2010; Schilbach, et al., 2012; Sebanz, et al., 2006).

If reciprocal relationships are important for joint actions, the nature of these reciprocal relationships (e.g. the social context) should influence the mechanisms underlying joint action. I implemented a novel experimental environment that takes into account the social context of joint action in Study 3. This "closed loop" environment allowed pairs of participants to play table tennis in a dark room with only the ball, net, and table visible. Visual information about both players' actions was manipulated by means of self-glowing markers that were selectively attached to the player's body or paddle. This allowed for assessing the importance of different sources of visual information on participant's performance when pairs of participants engaged in a table tennis game. Contrasting to the VE, the "closed loop" paradigm allowed for reciprocal interactions between individuals over the time course of the table tennis game. For instance, the "closed loop" paradigm allowed a player to perform a set of successive table tennis strokes in order to outperform the interaction partner (in a competitive context) or in order to keep the ball on the table (in a cooperative context). Hence, the "closed loop" paradigm allowed for treating the social context as an independent variable of experimentation, as suggested by several leading scientists in the field (Becchio, et al., 2010; Schilbach, et al., 2012; Sebanz, et al., 2006).

Study 3 employed the "closed loop" paradigm to test the role of visual information about an interaction partner's body and paddle on participant's performance (in terms of hit rate) in different social contexts. Specifically, we examined whether the instruction to play table tennis competitively or cooperatively affected the kind of visual cues necessary for successful table tennis performance. In two experiments, participants played table tennis in a dark room with only the ball, net, and table visible. Visual information about both players' actions was manipulated by means of self-glowing markers. We recorded the number of successful passes for each player individually. The results showed that participants' performance increased when their own body was rendered visible in both the cooperative and the competitive condition. However, social context modulated the importance of different sources of visual information about the other player. In the cooperative condition, seeing the other player's paddle had the largest effects on performance increase, whereas in the competitive condition, seeing the other player's body resulted in the largest performance increase. These results suggest that social context modulates the use of visual information. Further, the results suggest that a realistic assessment of the critical sources of visual information requires the consideration of the social context within which a joint action occurs.

Declaration of the Contribution of the Candidate

This thesis is presented in the form of a collection of manuscripts that are, at the time of thesis submission, either published or prepared for publication. The bibliographic details of the studies and where they appear in the thesis are set out below, together with a description of the contribution of each author.

Study 1

Streuber, S., Mohler, B.J., Bülthoff, H.H., de la Rosa, S. (2012). The Influence of Visual Information on the Motor Control of Table Tennis Strokes. *Presence: Teleoperators and Virtual Environments* (in press)

The idea for this study was proposed by the candidate. Design, stimulus generation, experimental work and analysis of the study have predominantly been developed and finalized by the candidate. The work has been presented at international conferences by the candidate. The co-author's role was that of supervision in giving advice, offering knowledge and criticism, and revising the manuscript.

Study 2

Streuber, S., Mohler, B.J., Bülthoff, H.H., de la Rosa, S. The Role of Predicting Temporally Proximal and Distal Events in Order to Interact with an Object Within a Social Setting *(in preparation)*

The idea for this study was proposed by the candidate. Design, stimulus generation, experimental work and analysis of the study have predominantly been developed and finalized by the candidate. The co-author's role was that of supervision in giving advice, offering knowledge and criticism, and revising the manuscript.

Study 3

Streuber, S., Knoblich, G., Sebanz, N., Bülthoff, H. and de la Rosa, S. (2011). The Effect of Social Context on the Use of Visual Information. *Experimental Brain Research* 214(2) 273-284.

The idea for this study was proposed by the candidate. Design, stimulus generation, experimental work and analysis of the study have predominantly been developed and finalized by the candidate. The work has been presented at several international conferences by the candidate. The co-author's role was that of supervision in giving advice, offering knowledge and criticism, and revising the manuscript.

Chapter 1: The Influence of Visual Information on the Motor Control of Table Tennis Strokes

The current chapter was published:

Streuber, S., Mohler, B. J., Bülthoff, H. H., & de la Rosa, S. (2012). The Influence of Visual Information on the Motor Control of Table Tennis Strokes. *PRESENCE: Teleoperators and Virtual Environments, 21*(3), 281-294.

Introduction

When humans interact with the world, they coordinate their body movements in real-time based on the sensory information about their environment (Bootsma & Vanwieringen, 1990; Grierson, et al., 2009; McBeath, et al., 1995; McLeod & Dienes, 1993a; Sarlegna & Blouin, 2010). For example, baseball players adjust their catching behavior to disturbances of the baseball's flying trajectory induced while they are catching the ball (Fink, et al., 2009). Other evidence for real-time action coordination comes from the observation that humans adjust their arm movements to disturbances of the object position induced while they are grasping an object (Reichenbach, et al., 2009). Moreover control laws describe how the central nervous system continuously converts sensory inputs into motor outputs (see, for example, reviews by Turvey, 1990; Warren, 2006). These studies suggest a close link between visual information and the online control of motor movements when humans interact with physical objects.

In everyday life situations humans do not exclusively interact with physical objects (object interaction), but very often they interact also with other humans (social interactions), e.g., handing over objects, playing games or carrying a sofa together. The effect of visual information on motor control has been systematically investigated in object interaction tasks (Reichenbach, et al., 2009). However, relatively little is known about the effect of visual information on motor control in social interaction tasks (Georgiou, Becchio, Glover, & Castiello, 2007). An investigation into the relevant sources of visual information for social interaction tasks would provide important insights into the mechanisms underlying human's ability to anticipate observed actions in order to facilitate the performance of social interactions.

Previous research provides evidence that humans are able to use visual information about another person to improve their performance in various tasks. For example, seeing the opponents body or interaction tool (i.e. a paddle or racket) improved ball prediction performance in table tennis (Streuber, Knoblich, Sebanz, Bülthoff, & de la Rosa, 2011), tennis (Huys, et al., 2009; Mann, Abernethy, & Farrow, 2010), squash (Abernethy, 1990), soccer (Savelsbergh, et al., 2002), and basketball (Aglioti, et al., 2008; Sebanz & Shiffrar, 2009). Most of the previous research found that experts (as compared to novices) are superior in predicting the fate of observed actions (Aglioti, et al., 2008; Calvo-Merino, Glaser, Grezes, Passingham, & Haggard, 2005; Casile & Giese, 2006; Keller, Knoblich, & Repp, 2007) although non-experts have the ability to understand the actor's expectations (Runeson & Frykholm, 1983), and intentions (Barrett, et al., 2005; J. Grezes, C. D. Frith, & R. E. Passingham, 2004b). Therefore, one might hypothesize that non-experts behavior in social interactions might not be affected by visual information about another person.

However, some research suggests that visual information about another person influences non-experts movement kinematics in a social interaction task. We define task performance as a measurement of human behavior in a given task at a specific point in time (e.g., the accuracy with which a participant gives a certain judgment at the end of an experimental trial) while movement kinematics are spatial-temporal measures of human limb movements (e.g., the velocity profile of an action). For example, Schmidt, Carello, & Turvey (1990) showed that two participants who were instructed to do cycling leg movements in an out-of-phase manner relative to each other, suddenly shifted to an in-phase leg movement. Since participants had only visual information about the other person available, these results suggest that visual information about the other person affects movement kinematics in naïve participants. Overall this research suggests that visual information about another person affects movement kinematics but less so the performance of non-experts (Schmidt, Carello, & Turvey, 1990).

To our knowledge, the effect of visual information on movement kinematics and task performance has not yet been compared within the same task. The result that visual information about another person affects non-experts' movement kinematics but not their task performance could also be attributed to other factors, i.e. varying task difficulties across studies. Therefore, in the present study we sought to examine the effect of visual information about another person on non-expert's movement kinematics and task performance within the same task.

The examination to what degree visual information about another person translates into differences in movement kinematics and task performance aids the understanding of how visual information and motor control are linked when non-experts perform social interaction tasks. This present study goes beyond previous research in three important aspects. First, we compare the effect of visual information on movement kinematics and task performance within the same task. Previous studies mainly focused on the investigation of either movement kinematics or task performance (Bideau et al., 2004; Craig, Berton, Rao, Fernandez, & Bootsma, 2006; Fink, et al., 2009; Schmidt, et al., 1990). An examination of how movement kinematics and task performance are altered by visual information requires that both are examined within the same task. Second, we are interested in the effect of visual information on task performance and movement kinematics under high fidelity realistic social interaction conditions. Hence unlike previous studies on movement kinematics that examined the rhythmic movements of participants, the present task did not involve cyclic movement but a more natural social interaction task, namely table tennis. Third, we varied the availability of different sources of visual information. Previous research suggests that not all sources of visual information are equally effective for social interaction performance (Streuber, et al., 2011). Hence we varied the availability of different sources of visual information about the interaction partner and the ball, in order to determine the visual variables important for motor control and consequently for social interaction performance.

In order to systematically investigate the effect of visual information on movement kinematics and task performance we designed a virtual reality table tennis experiment in which participants were asked to respond (for more information see below) to table tennis strokes performed by a virtual table tennis opponent. The virtual table tennis opponent was a computer generated avatar who randomly executed one out of twelve different table tennis strokes, which were pre-recorded with a motion capture system and played back as animations to the participant via a head mounted display (HMD). The experiment was conducted in an immersive interactive virtual environment using a HMD and motion tracking of the participant's head and table tennis paddle. The participant's task was to hit the virtual ball stroked by the virtual opponent as natural and accurate as possible while visual information about the ball, the paddle and the body of the opponent player was manipulated. This setup allowed us to provide a close-to natural interaction of the participant with his/her environment while at the same time it gave us full control over the manipulation of the visual information available to the participant. Additionally, the motion tracking of the table tennis paddle allowed an analysis of participants' movement kinematics and task performance. Importantly we applied all manipulations of the visibility of the ball, the paddle, and the body of the virtual opponent to all twelve pre-recorded animations in exactly the same way. As a result observed differences between the different visibility manipulations cannot be attributed to specific movement patterns or other characteristics of the virtual opponent. We used task performance and movement kinematics as dependent variables as derived from the motion capture data of the participants' paddle.

Task performance was defined as the minimum distance between the center of the paddle and the center of the table tennis ball throughout the movement trajectory of a participant's table tennis stroke (hereafter simply referred to as radial error). Note that the link between radial error and hitting performance is given by the fact that a ball would be missed (e.g., not hit) if the radial error exceeds the size of the table tennis paddle. We preferred radial error over a binary coding of performance (e.g., hit vs. non hit) because radial error is a continuous and therefore a more sensitive measure of performance than a binary measure. Radial error was used in previous studies in order to measure behavior in interceptive sports (N. Vignais et al., 2010).

Movement kinematics was measured in terms of stroke speed variability. Stroke speed variability was the variability of the speed profiles (speed over time throughout the stroke) of repeatedly executed strokes (see design section for a detailed description how stroke speed variability was derived from the 3D recordings of the participant's paddle). We choose stroke speed variability over the actual 3D movement trajectory as a movement kinematics measure for the following reason. 3D movement trajectories were expected to be very noisy in our experiment since the 12 pre-recorded ball trajectories required participants to conduct quite different stroke patterns due to directional changes of the strokes (e.g., striking to the left or right when the ball flew to left or right side of the participant respectively). This variability of the stroke patterns would have added additional noise to the data during the analysis therefore reducing power of the statistical analysis. Note that the first temporal derivative of the movement trajectory and its associated variability (i.e. stroke speed variability) is less sensitive to the directional changes of the stroke patterns. We therefore chose stroke speed variability over 3D movement trajectories as a kinematic measure.

We hypothesized that if visual information about the other person influences task performance we would expect to find significantly different radial errors for conditions in which the ball, the table tennis paddle and the body of the virtual opponent is visible compared to invisible. Likewise if visual information about the other person or the ball has an effect on movement kinematics we would expect to find significantly different stroke speed variability between the conditions when the ball, the paddle and the other person's body are visible compared to invisible.

Method

Participants

Ten participants performed the experiment (mean age: 25.6; SD: 2.12; 5 females). All participants had normal or corrected-to-normal vision and were right-handed. Participants were recruited from the Max Planck Institute subject database and were naive with respect to the purpose of the study. This research was performed in accordance with the ethical standards specified by the 1964 Declaration of Helsinki. All participants gave their informed consent prior to the experiment and were paid (8 Euro/hour) for their participation.

Stimulus and Apparatus

Participants played table tennis within the Virtual Environment (VE). The VE consisted of a virtual table tennis table (standard size: 2.74 m long x 1.525 m wide x 0.76 m high, with a 15.25 cm high net in the middle) and 12 animations of table tennis strokes (see Figure 1 for the 12 different ball trajectories). The animations included a virtual table tennis player, holding a standard table tennis paddle (physical radius of ~8cm) and a virtual table tennis ball (40mm diameter). The movements of participants were tracked using 16 infrared cameras (Vicon MX-13, using Vicon Tracker 1.1 tracking software (120Hz)) that tracked 10 infrared-reflecting markers rigidly attached to the participant's head and to the participant's paddle. Participants saw a stereoscopic image of the VE

21

through the head mounted display (nVisor SX60) from an egocentric perspective. The nVisor SX60 has a vertical FOV of 35 and a horizontal FOV of 44 with a resolution of 1280x1024 per eye. Participants also saw a virtual representation of the paddle in their hand. Figure 3 shows one participant wearing the head mounted display and holding the paddle.

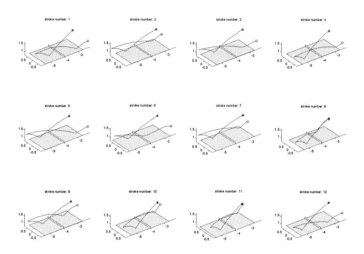

Figure 1: A visualization of the 12 different ball trajectories. The ball starts its trajectory from the participant's side (black filled dot on the right) towards the virtual player's side (left side). Then the virtual player returns the ball towards the participant who has to hit the ball. The black circled dot indicates the end of the ball trajectory. Six of the ball trajectories were played back hand (numbers 2, 3, 5, 6, 7, and 9) and the other six played front hand (numbers 1, 4, 8, 10, 11, and 12) by the virtual player. Additionally, six of the ball trajectories were played to the left side of the participant (2, 7, 8, 9, 10, and 11) and six trajectories were played to the right side of the participant (number 1, 3, 4, 5, 6, and 12).

Figure 2: A visualization of three frames of one of the 12 different table tennis stroke animations: The animations (left) were created by using motion capture data from an actor performing table tennis strokes (right). All 12 strokes were motion captured using a VICON motion capture system with 16 infrared cameras. The cameras recorded 46 (24 for upper body and 22 for lower body) infrared markers which were attached to the body of the actor, five markers which were attached to the paddle and the table tennis ball was painted with material that reflects infrared light.

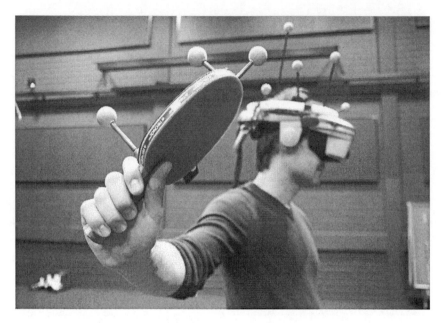

Figure 3: Photograph of a participant holding the paddle and wearing the head mounted display: Participants had to hit the virtual table tennis balls seen as a stereoscopic image within the head mounted display. The motion trajectory of the paddle was recorded and analyzed.

Animations

The animations were recorded from the body movements of an actor performing table tennis strokes using motion capture technology (see Figure 2). The actor was an athletic female (1.66 m, 20 years old, Division I college athlete) with no major table tennis experience. In order to obtain the animations the actor was instructed to perform 12 different table tennis strokes. The actor was holding a paddle and was standing at the end of the table. A second person (which was not recorded by motion capture) was placed on the opposite side of the table. The second person threw table tennis balls to the actor's side of the table. Six balls were thrown to the right and six balls to the left side of the table. If the ball was thrown on the right side, the actor responded with a forehand stroke, otherwise with a backhand stroke. Half of the balls the actor stroked to

24

the right and the other half to the left side of the opposite side of the table. All 12 strokes were motion captured using the motion capture system. The cameras recorded 46 (24 for upper body and 22 for lower body) infrared markers which were attached to the body of the actor, 5 markers which were attached to the paddle, and the table tennis ball was painted with material that reflects infrared light. The painting of the table tennis ball had no noticeable effects on its physical properties. The motion capture data were post-processed using the Vicon IQ 2.5 software. Missing parts of the ball trajectories (120fps) were fixed using a Vicon IQ 2.5 kinematic fitting filter (less than 1% of the frames were missing). Furthermore markers on the actor and paddle were automatically labeled by the Vicon objects (actor, and paddle) and hand-corrected when necessary. The skeletons (vsk-files) and the animations (v-file) were exported. These files were imported into the Autodesk MAYA animation software (we used the code provided by Carnegie Mellon University for import: http://mocap.cs.cmu.edu/tools.php). Finally these skeletons were attached to a character model (Complete Character set © Rockctbox Studios GmbH), to a ball model, and to a model of a paddle. Finally the whole animation was exported to Virtools 4.1 (using the Virtools exporter from Dassault Systems). These steps were repeated for all 12 stroke animations. Figure 2 shows the twelve different ball trajectories for the twelve different stroke animations used for all experimental conditions.

Design

The effect of visibility on movement kinematics and task performance was investigated in cight experimental conditions as outlined in Figure 4. We manipulated visual information about the virtual player's body (visible vs. invisible), the virtual player's paddle (visible vs. invisible), and the ball (visible vs. invisible) in a fully crossed 2x2x2 factorial design. The dependent variables measured movement kinematics and task performance. Task performance was measured in terms of radial error. Radial error was defined as the minimum distance between the center of the paddle and the center of the ball throughout the stroke. Movement kinematics was measured in terms of the variability in speed of repeatedly executed strokes. Speed profiles (speed over time throughout the stroke) were calculated for each stroke from the 3D motion capture recordings of the participant's paddle (center of the paddle) (see Figure 6(a)). The standard deviation profiles from the mean speed profiles were calculated for each

25

condition and participant (see Figure 6(b)). Finally, the standard deviations from the mean were integrated over time (from the moment the animation started to the moment the participant hit the ball). The sum of standard deviations is taken as a measurement of how consistently strokes were executed. Note that if a person would hit the ball with exactly the same speed profile then the stroke speed variability would be zero. If the ball would intercept the center of the paddle, radial error would be zero.

Procedure

Participants were placed in the middle of the tracking hall and were equipped with an HMD and a paddle. Participants were instructed to hit the ball that was served by the virtual player as natural and accurate as possible. Participants were not instructed to hit the ball back to the virtual player. Before the experiment started participants completed a practice block which familiarized participants with the VE equipment and the task. In this practice block participants hit virtual static balls that were presented at eight different locations above the participant's end of the table. If a ball was hit a sound was played, the ball disappeared and the next ball appeared at a new location. After successfully hitting eight balls (with a radial error less than 10cm) the exercise was completed and the experiment started.

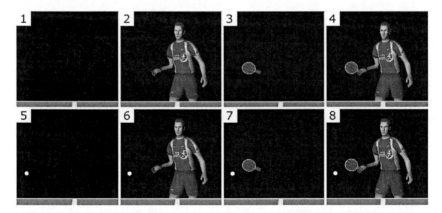

Figure 4: Visual stimulus for the eight different experimental conditions: We manipulated the visibility of the ball (visible vs. invisible), the paddle (visible vs. invisible), and the body of the virtual player (visible vs. invisible).

Each participant played 4 experimental blocks. Each experimental block consisted of the presentation of the 12 animations under each of the 8 experimental conditions (=96 trials). The order of the 96 trials in each block was randomized. Each trial was 2800ms long and started with a 400ms long beep sound to indicate the start of the trial. Afterwards one out of 12 animations started playing. In each animation the ball started flying from the participant's side of the table towards the virtual player's side of the table. Then the virtual player stroked the ball back to the participant who attempted to hit the ball. A visual (text: "HIT" which was placed in the center of the screen for 1 second) and acoustic feedback (500ms sound) informed the participant when the radial error was below 50cm. There was no visual or acoustic feedback if the ball was missed. The feedback was purely motivational. Importantly, all strokes were used for data analysis whether the feedback was provided or not. In order to continue to the next trial participants were required to move back into the initial body posture (standing in the center with the right arm relaxed so that the paddle was positioned in parallel to the right thigh).

Results

Radial error:

First, in order to see whether performance changed with time (e.g., due to learning, unlearning, or fatigue, etc.) an ANOVA was conducted on participants mean radial error scores with the four repetitive blocks as a factor. The ANOVA revealed no significant effect of block on radial error, $F_{(9,3)}=1.77$; $p=0.176$. Thus, there was no significant learning effect over time and further analysis of mean radial error was collapsed over blocks.

We tested the effect of visibility of visual information on radial error using a completely crossed within subject ANOVA with visibility of the ball (visible vs. invisible), visibility of the paddle of the virtual player (visible vs. invisible) and visibility of the body of the virtual player (visible vs. invisible) as factors.

The ANOVA revealed a significant main effect of ball, $F_{(1,9)}=200.22$, $\eta^2_{partial}=0.957$, $p<0.001$, a significant main effect of paddle, $F_{(1,9)}=7.42$, $\eta^2_{partial}=0.452$, $p=0.023$, and a significant main effect of body, $F(1,9) =14.31$, $\eta^2_{partial}=0.614$, p=0.004. There was a significant interaction between ball and body, $F(1,9) =15.00$, $\eta^2_{partial}=0.625$, p=0.004, and a significant interaction between paddle and body, $F(1,9) =8.04$, $\eta^2_{partial}=0.472$, p=0.02. The interaction between ball and paddle was not significant, $F(1,9) =4.39$, $\eta^2_{partial}=0.328$, p=0.066. There was also a significant three-way interaction between ball, paddle and body, $F(1,9) =10.95$, $\eta^2_{partial}=0.549$, p=0.009. The three-way interaction confirms the statistical significance of the above observation that the ball visible and ball invisible condition differ with respect to how the factors paddle and body interact (see Figure 7(a) and 7(b)). We dissected the significant three way interaction by calculating two separate two-way within subject ANOVAs: one for the ball visible and one for the ball invisible condition.

The first ANOVA was run on the subset of the data where the ball was invisible. The ANOVA revealed a significant main effect of paddle, $F(1,9) =7.41$, $\eta^2_{partial}=0.452$, p=0.024 and body, $F(1,9) =15.86$, $\eta^2_{partial}=0.638$, p=0.003. The significant interaction between paddle and body, $F(1,9) =9.61$, $\eta^2_{partial}=0.516$, p=0.013, is shown in Figure 7(a). Paired t-tests were used to compare the effect of seeing the paddle on radial error depending on whether the body was visible. Seeing the paddle decreased the radial error only if the body was invisible, t(10)=3.23, p=0.010, but not if the body was visible, t(10)=0.69, p=0.505. In sum, if the ball was not visible, any information (body and/or paddle information) about the virtual player was helpful in reducing the radial error.

The second ANOVA was run on the data of the ball visible conditions. The ANOVA was conducted to test if there are significant differences in the radial error depending on the visibility of the paddle and the body of the virtual player. The ANOVA revealed no significant effect of paddle, $F(1,9) =2.27$, $\eta^2_{partial}=0.002$, p=0.166, no significant effect of body , $F(1,9) =0.02$, $\eta^2_{partial}=0.452$, p=0.894, and no significant interaction between paddle and body, $F(1,9) =0.06$, $\eta^2_{partial}=0.007$, p=0.813 when the ball was visible.

In sum, the radial error was always lower when the ball was visible (mean=0.234m; SD=0.047) compared to invisible (mean=0.537m; SD=0.062). Moreover, the visual information about the virtual player's body and/or the paddle helped only when the ball was not visible.

Stroke speed variability (SSV):

We examined the effect of visual information on SSV in an ANOVA with the within-subject factors: visibility of the ball (visible vs. invisible), visibility of the paddle of the virtual player (visible vs. invisible) and visibility of the body of the virtual player (visible vs. invisible).

The ANOVA revealed a significant main effect of paddle, $F(1,9) = 28.47$, $\eta^2_{partial} = 0.760$, $p < 0.001$, and a significant main effect of body, $F(1,9) = 21.05$, $\eta^2_{partial} = 0.701$, $p = 0.001$. However there was no significant effect of ball, $F(1,9) = 0.86$, $\eta^2_{partial} = 0.087$, $p = 0.379$. There was also a significant interaction between paddle and body, $F(1,9) = 26.09$, $\eta^2_{partial} = 0.744$, $p = 0.001$. However there was no significant interaction between ball and paddle, $F(1,9) = 4.23$, $\eta^2_{partial} = 0.320$, $p = 0.070$. Also the interaction between ball and body was not significant, $F(1,9) = 2.27$, $\eta^2_{partial} = 0.202$, $p = 0.166$. Finally there was no significant three-way interaction between ball, paddle and body, $F(1,9) = 2.52$, $\eta^2_{partial} = 0.218$, $p = 0.147$. The significant interaction between paddle and body is shown in Figure 7(c) and Figure 7(d). Paired t-tests were used to compare the effect of seeing the body on SSV depending on whether the paddle was visible or not. Seeing the body decreased SSV if the paddle was invisible, $t(10) = 4.92$, $p = 0.001$, but also if the paddle was visible, $t(10) = 2.49$, $p = 0.034$. However, seeing the paddle only decreased SSV when the body was invisible, $t(10) = 5.41$, $p < 0.001$ but not when the body was visible, $t(10) = 1.40$, $p = 0.195$. In sum these results indicate that seeing the body of the virtual player always decreased SSV. Seeing the paddle of the virtual player only decreased SSV when the body of the virtual player was not visible. However, the visibility of the ball did not affect SSV. Note that these results complement the results from the analysis of the radial error data. While the radial error data showed that the visibility of the ball leads to the best task performance (in terms of a low radial error), the variability data suggest that movements are carried out in a more consistent manner (in terms of low SSV) if the body is visible. Hence our analysis of movement kinematics revealed additional aspects that would have gone unnoticed by assessing task performance only.

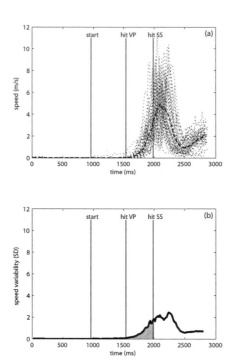

Figure 5: A visualization of the calculation of stroke speed variability. The 3D trajectory of each stroke (center of the participant's paddle) was recorded with the motion capture system. Panel (a) shows the 48 speed profiles (thin black scattered lines) and the mean speed profile (black thick scattered line) of one participant in one viewing condition (condition 8, body, paddle and ball visible) over the time course of the stroke. This 48 speed profiles where derived from 12 stroke repetitions and 4 block repetitions. The black thick line in panel (b) indicates the standard deviation from the mean speed profile (see panel (a)) over the time course of the stroke (stroke speed variability profile). The left vertical black line (start) indicates the mean point in time when the ball started its trajectory towards the virtual player. The middle vertical black line (VP) indicates the mean time when the virtual player hit the ball. The right black vertical line (SS) indicates the mean time when the participant hit the ball. The overall stroke speed variability was derived by integrating the stroke speed variability profile from the time when the ball started the trajectory (start) to the time when the participant hit the ball (SS).

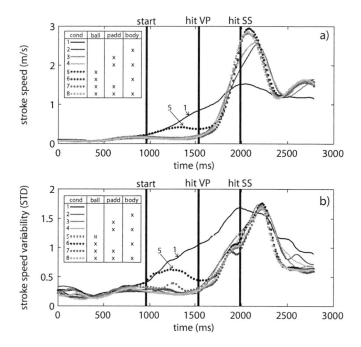

Figure 6: Stroke speed profiles (panel (a)) and stroke speed variability profiles (panel b). Panel a) shows the average speed profiles of participants' paddle movement over time for each condition. Panel (b) shows the average standard deviation from the mean stroke speed profiles (stroke speed variability) over time for each condition. Start indicates the mean time when the ball starts moving. Hit VP indicates the mean time when the virtual player hits the ball. Hit SS indicates the mean time when the participant hit the ball. Panel (b) shows that stroke speed variability was similar in all conditions, except in viewing condition one (where no visual information was available) and in the viewing condition five (where only the ball was visible). In those conditions stroke speed variability was significantly increased. Interestingly, these differences in stroke speed variability occurred already before the virtual player hit the ball (hit VP). The increase in stroke speed variability in condition one remains until the participant hits the ball (hit SS). However, the increase in stroke speed variability in condition five disappears after the virtual player hit the ball. In order to do the statistical analysis the overall stroke speed variability was calculated for each participant and condition by integrating stroke speed variability over time (from start to hit SS).

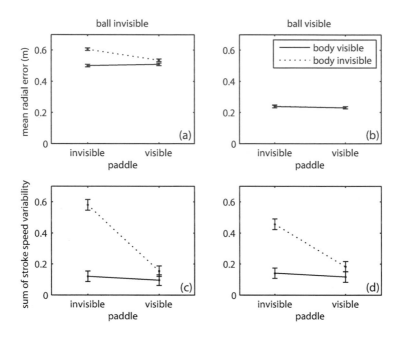

Figure 7: These graphs show the mean radial error (panel (a) and (b)) and mean stroke speed variability (panel (c) and (d)) coded with respect to the factors ball, paddle, and body. The error bars indicate the standard error from the mean as derived from the error term of the 3-way ANOVA. Panel (a) shows the significant interaction between paddle and body when the ball was invisible. Panel (b) shows no effect of paddle and body when the ball was visible. Panel (c) and panel (d) show a significant interaction between paddle and body that was not modulated by the visibility of the ball.

Discussion

This study investigated the role of different sources of visual information on error rate (task performance) and stroke speed variability (movement kinematics). Participants' task was to hit a virtual ball that was played by a virtual table tennis player as accurate and natural as possible. We found that radial error mainly relied on the visibility of the ball. Furthermore, we found that visual information about the virtual player's body and/or paddle reduced radial error when the ball was invisible. This result is explained by the fact that the ball provides task relevant information (participants task was to hit the ball). However, visual body and paddle information also provides task relevant information if the ball is not visible. This suggests that when people are forced to use other visual information to estimate the virtual ball location at the time of stroke, they can do so. Hence our results extend previous findings that showed that visual information about the other person's body does not induce any performance changes (N. Vignais, et al., 2010). Vignais and colleagues (2010) found that the level of graphical detail of a virtual handball thrower did not influence goal keeper's motor response. In their experiment, handball goal keepers were asked to stop a ball thrown by a virtual hand ball thrower with different levels of details: a textured reference level (L0), a non-textured level (L1), a wire-frame level (L2), a point-light-display level (L3), and a point-light-display level where the ball was also a point light display (L4). Performance was measured in terms of radial error, which was the minimal distance between the center of the goalkeepers hand and the center of the ball when the goalkeeper caught the virtual ball. The authors did not find significant differences in the radial error depending on the level of detail of the virtual thrower (L0 vs. L1 vs. L2 vs. L3), but significant differences depending on the level of detail of the ball (L0, L1, L2, L3 vs. L4). Note however, that in all conditions, the ball was visible. Our result extend these findings by showing that participants are also able to estimate the ball location from the body and paddle cues if the most task relevant information (the ball) is not provided.

Our finding that naïve participants can use body and paddle information to improve performance is relevant for theories of action understanding. Previous research suggests that expert players seem to be superior than naïve participants in improving their task performance in the presence of visual information about the other person's body (Abernethy, 1990; Aglioti, et al., 2008; Huys, et al., 2009; Mann, et al., 2010; Savelsbergh, et al., 2002; Sebanz & Shiffrar, 2009). This research is less clear about whether naïve participants can use visual information about the other person's body at all. Our finding that naïve participants improved their task performance (i.e. reduced the radial error) when only the other player's body or paddle was visible suggests that also non-expert players have the ability to anticipate the actions of an opponent if they are forced to do so (i.e. in the absence of visual cues about the ball).

The sensitivity of naive players to the other person's body and paddle information in our experiment is also indicated by the analysis of the movement kinematics in terms of stroke speed variability. A visual comparison of the average speed profiles suggests that participants started to move earlier when the virtual player was not visible (see Figure 6(a)). The visual inspection of the time courses further suggests that differences in stroke speed variability were largest just before the virtual player hit the ball (see Figure 6(b)). If the body of the virtual player was visible, stroke speed profiles were less variable. Seeing the paddle decreased variability only if the body was invisible. Finally, seeing the ball did not affect stroke speed variability. These results suggest that participants' movements were mainly guided by visual information about the virtual player's body. This result shows for the first time that motor movements in non-expert table tennis players are coupled to visual body information about an opponent. The behavioral differences were largest before and in the moment the opponent hit the ball, suggesting that participant's behavior was not purely affected by the visual presence of the virtual character but by the action the virtual character performed (striking the ball).

One possibility to explain reduced variability in the movement kinematics is by means of spontaneous synchronization which is a form of spontaneous pattern formation that operates according to general principles of self-organization as described by non-linear systems. Synchronization of body movements between individuals occurs unintentionally as soon as individuals share a medium of communication (e.g., visual body information about the other person)(Winfree, 2002). For instance, when performing rhythmic body movements in the presence of others, humans tend to synchronize phase and frequency of their movements with other persons without instruction to do so (Kelso, 1984; Lagarde & Kelso, 2006). Synchronization of the phase and frequency of limb movements has been observed for postural sway (Shockley,

34

Santana, & Fowler, 2003), swinging of hand-held pendulums (Schmidt, et al., 1990), finger tipping (Oullier, De Guzman, Jantzen, Lagarde, & Kelso, 2008), swinging legs (Kelso, 1984), rocking chairs (Richardson, Marsh, Isenhower, Goodman, & Schmidt, 2007), clapping (Neda, Ravasz, Brechet, Vicsek, & Barabasi, 2000), and when engaging in a verbal problem solving task (Richardson, Marsh, & Schmidt, 2005). Synchronization might also be able to explain the stroke variability effects. The visibility of another human should lead to larger synchronization and therefore to more consistent and less variable behavior. Therefore, if participants did synchronize their body movements to the body movements in the conditions in which the virtual player was visible, we would expect that the variance in the participant's movement kinematics decreases in these conditions. Hence interpersonal synchronization might be one explanation for the decreased stroke variability in the conditions in which the other player's body or paddle was visible.

Our results also support the long standing idea that action and perception are closely linked (Bootsma & Vanwieringen, 1990; Grierson, et al., 2009; McBeath, et al., 1995; McLeod & Dienes, 1993a; Sarlegna & Blouin, 2010). Here, we can only speculate about the mechanisms which led to a coupling between visual information about the virtual player and the participant's movement kinematics and task performance. In general different mechanisms can lead to spontaneous synchronization (see (Pikovsky, Rosenblum, & Kurths, 2001), for review). In the case of human coordination, spontaneous synchronization might be the result of a link between action observation and action execution. For instance, common coding theory (Prinz, 1984; Prinz & Hommel, 2002) suggests that there is a shared representation (a common code) for both perception (e.g., seeing an action) and action (performing an action). Referring to common coding, seeing an action activates the motor representation associated with this action. Common coding is also supported by neurobiological studies which identified neurons (so called mirror neurons) which both fire when an action is performed (e.g., grasping an object) and when the same action is observed (e.g., somebody else grasps an object) (Gallese, Fadiga, Fogassi, & Rizzolatti, 1996; Iacoboni et al., 1999). The core idea behind these theories is that common coding allows an observer to understand an observed action by means of sensory-motor representations. Action understanding might allow an observer to predict the outcome

35

of an observed action which is regarded as important for joint action coordination (Sebanz, et al., 2006). Even though relatively little is known about the functional role of common coding for interpersonal coordination, a common assumption is the observation of an interaction partner guides one's own movement kinematics and thus might support synchronization.

What might be the functional role of synchronization in a task like table tennis? It has been suggested keeping together in time and space is one of the most powerful ways to produce and reproduce communication (McNeill, 1995). Furthermore, asynchronous movements may be energetically more costly for the dyad than synchronous movements (see, e.g., (Kording, Fukunaga, Hovard, Ingram, & Wolpert, 2004). In addition, Sebanz et al. (2006) suggested that a coupling between action perception and action execution has a functional role for joint action performance. They argue that joint action performance relies on humans' ability to represent and predict others' actions and to integrate predicted effects of own and others action into one's own action planning. With this hypothesis a coupling between motor movements and visual information about the virtual player might lead to more consistent movement kinematics when the opponent is visible.

This study for the first time established a link between movement kinematics and task performance in a sensory motor coordination task. The results showed that non-expert table tennis players' behavior relies on the visibility of the ball in natural conditions. However, non-expert players are also able to use the visibility of the opponent's body and paddle to improve performance if visual information about the ball is not available. We further showed that variability of movement kinematics is affected by visual information about another person in non-expert players. Stroke speed variability was reduced when the opponent's body or paddle was visible. Overall our findings demonstrate that humans use visual information about the opponent in order to predict the ball trajectory and therefore support predictions in favor of current theories of joint action, i.e. common coding theory, which suggest a link between action observation and execution.

Acknowledgments

We would like to thank Erin McManus for helping with the motion capture of the table tennis strokes and Joachim Tesch for technical support. We thank Trevor Dodds and Eva Hanau for helpful comments on the manuscript. We gratefully acknowledge the support of the Max Planck Society and the WCU (World Class University) program through the National Research Foundation of Korea funded by the Ministry of Education, Science and Technology (R31-10008). The contribution of Stephan de la Rosa was funded by the EU Project TANGO (ICT-2009-C 249858).

Chapter 1: The Influence of Visual Information on the Motor Control of Table Tennis Strokes

Chapter 2: Evidence for Two Types of Action Predictions in More Natural Social Interactions

The current chapter is based on the draft: "The Role of Predicting Temporally Proximal and Distal Events in Order to Interact With an Object Within a Social Setting " by Stephan Streuber (the candidate of the current PhD thesis), B.J. Mohler, H.H. Bülthoff , and S. de la Rosa.

Introduction

Many everyday life situations require humans to interact with an object in some social setting, e.g. when playing table tennis or carrying a sofa. Although these tasks are highly complex and require the anticipation of future states of both the object and the interaction partner, humans often have little difficulties to coordinate their actions in these tasks. How do humans achieve this remarkable feat? Previous research on online control of motor movements in object interaction tasks suggests that visual information about the object is important. On the other hand research on action coordination and action understanding highlights the importance of visual information about the interaction partner in action coordination tasks. To date it is largely unknown how these two sources of visual information interact in an object interaction task that is embedded within a social setting. Understanding how people use different sources of visual information in an object interaction task that involves two people shed some light on the underlying psychological processes and advances our understanding of action coordination in more natural settings. The aim of the present study is to examine how visual information of the object and the interaction partner affect an object interaction task that occurs in a social setting.

Research examining online motor control in object interaction tasks suggests that humans adjust their body movements in real-time based on the sensory information about their environment (Bootsma & Vanwieringen, 1990; Grierson, et al., 2009; McBeath, et al., 1995; McLeod & Dienes, 1993a; Sarlegna & Blouin, 2010). For

39

instance, when catching a baseball, the catcher uses visual information about the ball (e.g. the velocity of the projection of the ball on his/her retina) to continuously adjust his/her behavior (moving forward or backwards) when catching the flying ball (Fink, et al., 2009). Similar strategies are used in navigation on basis of optic flow information or when intercepting moving or static objects (Bootsma & Vanwieringen, 1990; Reichenbach, et al., 2009).

On the other hand, research on joint actions (actions in which two or more people coordinate their action in space and time) often implicitly assumes that visual information about the other person is important for a successful joint action. For example, a successful joint action requires the prediction of another person's action. Action prediction is important to coordinate one's own action in response to the other person's action in a timely fashion (Adolphs, 1999, 2003; Sebanz, et al., 2006). The ability of humans to predict the outcome of an action on basis of perceptual information about another person has been demonstrated in tennis (Huys, et al., 2009; Mann, et al., 2010), squash (Abernethy, 1990), soccer (Savelsbergh, et al., 2002), and basketball (Aglioti, et al., 2008). The prediction of the interaction partner's actions allows the preparation and/or execution of an adequate complementary action. This in turn facilitates joint action.

Based on the previous research on motor control and joint action one would predict that both visual information about an object and about the interaction partner are important in an object interaction task that is carried out by two persons. How do these two different sources of visual information interact in an object interaction task that occurs in a social setting? Because action coordination is important for object interaction and joint action, at least part of the answer can be found in how visual information affects action coordination. Here Vesper and colleagues recently proposed that action coordination occurs on two different time scales (Vesper, Butterfill, Knoblich, & Sebanz, 2010). In particular they proposed that motor control of one's actions operates on a fine temporal scale and serves the prediction of temporally proximal events while the prediction of another person's cognitive states (e.g. intention or goals) are used to make predictions about temporal more distal events. How does this proposal transfers to a task in which two agents interact with each other by means of an object (object interaction task in a social setting)? Research on motor control suggests that visual information about the object should be informative for the direct interaction with the object. That is, visual information should improve performance on a given task shortly before participants interact with the object, i.e. for temporally proximal events. On the other hand, other research indicates that visual information about the interaction partner is at least partly indicative of his/her cognitive states, in particular, his/her action intentions (Grezes, et al., 2004a). Knowing the intentions of the other player might facilitate performance because observers can predict what the other person is going to do next. For example, goal keepers can better predict the fate of a penalty kick when observing the body of the penalty kicker prior to ball contact (Savelsbergh, et al., 2002). Also, basketball players can better predict the fate of a basketball shot when observing the body of the shooter (Aglioti, et al., 2008) before the ball is released from the hand. These results suggest that visual information about the other player's body enhances the prediction of temporally distant events, which in turn should also affect the interaction with an object in a social setting.

Based on these considerations we made the following predictions for a table tennis task in which participants are required to hit a ball that is played by a virtual agent in a Virtual Reality setting. Because visual information about the virtual agent is informative about intended distant events (e.g. action intentions), the visual information about the virtual agent should help participants, for example, with the decision to which side of the table the virtual agent will play the ball. Previous research suggests that this might be possible already before the virtual agent hits the ball (Aglioti, et al., 2008; Savelsbergh, et al., 2002). On the other hand the visual information of the ball should provide important cues for the direct interaction with the ball (one's own motor control) and consequently for temporally proximate events. We therefore expect that visual information about the ball affects participants shortly before they hit the ball. Because the interaction of the virtual agent with the ball (and predictions made from this interaction) precedes participants' interaction with a ball, visual information about the other player information should affect participants' performance at an earlier point of time than visual information about the ball. Overall we predicted that different sources of visual information should be useful at different times during an object interaction task that is embedded in a social context. Specifically we predict that visual information about another person should affect participants playing performance at an earlier point of time than visual information about the ball.

The examination of how different sources of visual information (in particular visual information about the other person and the ball) affect performance in a social object interaction task is important to bring joint action and motor control research a step closer. Importantly by examining how different sources of visual information, which each field considers critical, affect performance within the same task we are able to identify the task aspects that are important in an object interaction task under a more natural setting. Hence the investigation of the effect of different sources of visual information on performance in an object interaction task that occurs in a social setting is interesting for both for the research on motor control and joint action and advances our knowledge how human coordinate their action under more natural settings.

We tested how different sources of visual information affect performance in a Virtual Reality table tennis task. We manipulated the presentation time of different sources of visual information. The advantage of a VR table tennis task is to have full control over the visibility of different sources of visual information (ball, virtual agent's paddle, virtual agent's body). Each trial consisted of pre-recorded animations showing a virtual agent playing a virtual ball towards the participant. The task of the participant was to hit the virtual ball with a standard table tennis paddle. The table tennis paddle of the participants was motion tracked using an optical tracking system. The 3D-movement trajectories of the participants' strokes were later used for data analysis. During the animation sequence we manipulated the presentation duration of different sources of visual information in order to see whether the use of one particular source of visual information precedes the use of another source of visual information. Presentation duration measured how long a particular source of visual information was visible from the beginning of a trial. In order to examine the time point when a particular source of visual information became important, we compared participant's performance of temporally adjacent presentation duration conditions. A significant change in performance between adjacent presentation duration conditions would indicate that participants used the presented visual information differently in the adjacent presentation conditions.

We reasoned that if participants use the visual information about the virtual agent to predict the agent's playing intentions (in particular the side to which the virtual agent will play the ball), visual information about the virtual agent should affect participants' before the virtual agent interacts with the ball. In contrast, if visual information about the ball is important for the participants' online control of table tennis strokes, we expect that this information should affect participants' performance shortly before participants hit the ball. Because in our experimental design the virtual agent always interacts with the ball before the participant, we expect that the visual information about the virtual agent should affect participants' performance earlier than visual information about the ball.

Figure 2: A visualization of three frames of one of the 4 different table tennis stroke animations: The animations (left) were created by using motion capture data from an actor performing table tennis strokes (right). All 4 strokes were motion captured using a VICON motion capture system with 16 infrared cameras. The cameras recorded 46 (24 for upper body and 22 for lower body) infrared markers which were attached to the body of the actor, five markers which were attached to the paddle and the table tennis ball was painted with material that reflects infrared light.

We intended to test our hypothesis that participants use the visual information about the virtual agent for predicting the playing direction of the virtual agent. We reasoned if participants use the virtual agent's visual information (in particular the initial posture information) to infer the virtual agent's stroking direction, then assigning an incorrect initial posture to a stroke should lead to an incorrect prediction of the agent's stroking direction. We therefore manipulated the validity of the initial body and paddle posture cues across different strokes in the following way. Note, usually diagonal and parallel table tennis strokes are associated with different player's body postures. We paired both diagonal and parallel strokes with an initial posture that is normally used for diagonal strokes. Hence the initial postural cue indicates the correct hitting direction for diagonal strokes (valid postural cue condition), but the opposite direction for parallel strokes (invalid postural cue condition). If participants use visual information about the virtual agent for predicting virtual agents hitting direction (i.e. intention), participants should be able to predict the correct direction in the valid postural cue condition and should predict the incorrect direction (opposite direction) in the invalid postural cue condition.

We designed a virtual reality table tennis experiment to examine the temporal dimension about the use of different sources of visual information. In particular we were interested in whether visual information about another person is used to infer temporally distant events (i.e. the playing direction of the ball). Moreover we set out to test the hypothesis that the use of visual information about another person occurs prior to the use of visual information about an object. In this experiment participants played table tennis with a virtual agent while the availability of visual information was manipulated in different time phases. The virtual agent was a computer generated avatar. Participants were equipped with a head mounted display (HMD) to see the virtual environment. Participant's task was to hit the virtual ball (stroked by the virtual agent) as naturally and accurately as possible with their table tennis paddle. This setup allowed us to provide a close-to natural interaction of the participant with his/her environment. At the same time the setup gave us full control over the manipulation of the visual information available to the participant. Additionally, the motion tracking of the participants table tennis paddle allowed an analysis of the 3D-trajectories for each table tennis stroke. From the analysis of the 3D-trajectories we derived our dependent variables as discussed in the method section. In sum this setup allowed us to access the effect of different sources of visual information on action coordination at different time phases of the social interaction.

Methods

Participants

Twenty one participants performed the experiment. All participants had normal or corrected-to-normal vision and were right-handed. Participants were recruited from the Max Planck Institute subject database and were naive with respect to the purpose of the study. This research was approved by the ethics committee of the University of Tuebingen and was carried out in accordance with the ethical standards specified by the 1964 Declaration of Helsinki. All participants were paid (8 Euro/hour) for their participation.

Stimulus and Apparatus.

Participants played table tennis within the Virtual Environment (VE). The VE consisted of a virtual table tennis table (standard size: 2.74 m long x 1.525 m wide x 0.76 m high) and a virtual agent who performed forehand or backhand strokes. The virtual agent stroked virtual table tennis balls (40mm diameter) to the left or right side of the table towards the participant. The movements of the participants were tracked using 16 infrared cameras (Vicon MX-13, using Vicon Tracker 1.1 tracking software (sampling rate 120Hz)) that tracked 10 infrared-reflecting markers rigidly attached to the participants head and to the participant's paddle. Participants saw a stereoscopic image of the VE through the head mounted display (nVisor SX60) from an egocentric perspective. The nVisor SX60 has a vertical FOV of 35 and a horizontal FOV of 44 with a resolution of 1280x1024 per eye. Participants also saw a virtual representation of the paddle in their hand. Figure 1 shows one participant wearing the head mounted display and holding the paddle.

Animations

The animations of the virtual player were retrieved from prerecorded body movements of an actor performing table tennis strokes using motion capture technology (see Figure

2). The actor was an athletic female (1.66 m, 20 years old, Division I college athlete) with no major table tennis experience. In order to obtain the animations the actor was instructed to perform 4 different table tennis strokes. The actor was holding a paddle and was standing at the end of the table. A second person (which was not recorded by motion capture) stood on in the middle of the opposite side of the table. The second person threw table tennis balls to the actor's side of the table one by one. Two balls were thrown to the right and two balls to the left side of the actor. If the ball was thrown on the right side, the actor responded with a forehand stroke, otherwise with a backhand stroke. The actor stroked half of the balls (for both forehand and backhand strokes) to the right and the other half to the left side of the thrower. The actor was instructed to keep the initial posture as if she were stroking to the opposite side of the table regardless of the side to which she was supposed to stroke the ball. Hence for two strokes the initial body posture matched the playing direction (valid posture cue condition) and for another two strokes (invalid posture cue condition) the initial body posture did not match the stroking direction. All 4 strokes were motion captured using the motion capture system. The cameras recorded 46 (24 for upper body and 22 for lower body) infrared markers which were attached to the body of the actor, 5 markers which were attached to the paddle, and the table tennis ball was painted with material that reflects infrared light. The painting of the table tennis ball had no noticeable effects on its physical properties. The motion capture data were post-processed using the Vicon IQ 2.5 software. Missing parts of the ball trajectories (120fps) were fixed using a Vicon IQ 2.5 kinematic fitting filter (less than 1% of the frames were missing). Furthermore markers on the actor and paddle were automatically labeled by the Vicon objects (actor, and paddle) and hand-corrected when necessary. The skeletons (vsk-files) and the animations (v-file) were imported into the Autodesk MAYA animation software (we used the code provided by Carnegie Mellon University for import: http://mocap.cs.cmu.edu/tools.php). Finally these skeletons were attached to a character model (Complete Character set © Rocketbox Studios GmbH), to a ball model, and to a model of a paddle. Finally the whole animation was exported to Virtools 4.1 (using the Virtools exporter from Dassault Systems). These steps were repeated for all stroke animations.

Design

We tested the effect of source of visual information, presentation duration, and posture cue validity on task performance in a fully crossed three-factorial design. The factor *source of visual information* coded four different viewing conditions: (1) only ball visible viewing condition; (2) only body visible viewing condition; (3) only paddle visible viewing condition; and (4) all visible viewing condition. An overview of the four different viewing conditions is provided in Figure 3. Factor *presentation time* coded how long visual information about a specific source of visual information was visible: (1) until one frame after the ball started flying in the direction of the virtual opponent; (2) until one frame after the ball hit the table the first time; (3) until one frame after the ball is hit by the virtual opponent; (4) until one frame after the ball hits the table a second time; (5) the animation sequence was played until the end. Figure 4 provides an overview of the 5 different presentation durations. Finally, the factor *posture cue validity* had two levels (valid and invalid).

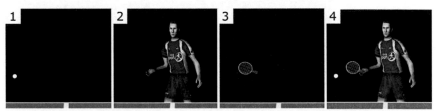

Figure 3: Visibility of different sources of visual information in the four different viewing conditions: We manipulated the visibility of the ball (visible vs. invisible), the paddle (visible vs. invisible), and the body of the virtual player (visible vs. invisible).

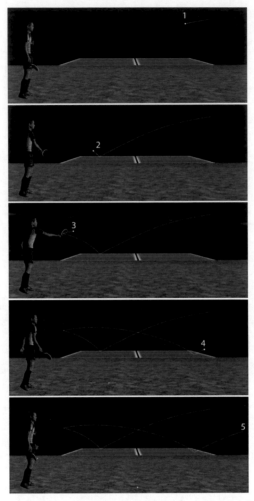

Figure 4: Factor presentation duration coded how long visual information about a specific source of visual information was visible: (Panel 1) until one frame after the ball started flying in the direction of the virtual opponent; (Panel 2) until one frame after the ball hit the table the first time; (Panel 3) until one frame after the ball is hit by the virtual opponent; (Panel 4) until one frame after the ball hits the table a second time; (Panel 5) the animation sequence was played until the end.

Task performance was measured in terms of the directional accuracy (DA). In order to calculate DA we first determined the estimated point of impact (EPI). EPI measured the point in the movement trajectory of the paddle where the participant thinks to hit the ball. A pilot analysis of 12 recorded table tennis strokes performed by a research assistant revealed that the maximum velocity of the paddle coincides with the moment at which the ball was hit with the paddle (see Figure 5). Hence the maximum velocity can be used as an indicator about a person's estimation about the point of impact of the ball onto the paddle. EPI was then used to calculate the dependent variables in the following way.

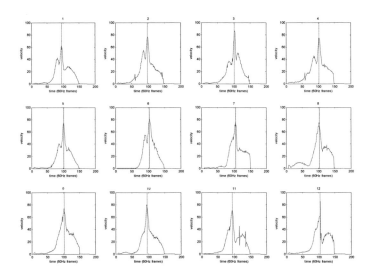

Figure 5: The different panels show the velocity profiles of the 12 3D-movement trajectories of a research assistants paddle when performing table tennis strokes (red lines). The blue line in each panel shows the time when the ball touched the paddle. We found that for all 12 strokes the time of the maximal velocity coincides with the time the ball touches the paddle. From this we concluded that the time of the maximum velocity in the movement trajectory of a paddle when performing a table tennis stroke can be used in order to infer when a participant intents to hit a ball.

Direction accuracy (DA) measured the percentage of trials in which the spatial location of the paddle at the EPI corresponded to the same side of the table (reference point table midline) on which the ball arrived. In other words DA measures the amount of trial on which participants executed the stroke on the same side of the table as the ball arrived. If participants randomly picked a side for executing a stroke DA should be equal to 50%. Likewise a DA of less than 50% indicates that participants executed more strokes to the opposite side (incorrect side) of the table than of the same side of the table where the ball arrived (correct side). We calculated the average DA for each condition separately. We reasoned that if a particular source of visual information is important for selecting the direction of the stroke, DA should lie significantly above 50% chance level in the corresponding viewing condition. On the other hand a DA value significantly smaller than 50% indicates that a particular source of visual information misleads participants to stroke to the opposite direction. Figure 7 shows a typical stroke trajectory including the EPI and the velocity profile.

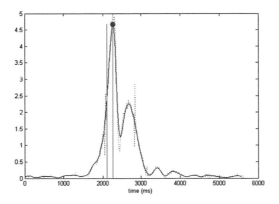

subject:1 block:1 trial:10 condition:3 time:2

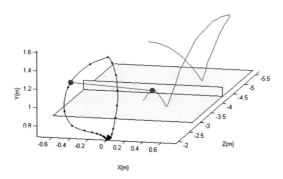

Figure 7: illustrates the calculation of the dependent variables as derived from the 3D-movement trajectory of the participants table tennis paddle and the ball flight trajectory. The blue line in the bottom panel shows a typical ball trajectory. The black line in the bottom panel shows a typical 3D-stroke trajectory as recorded from the participants paddle. The black line in the upper panel indicates the velocity profile of the participants paddle throughout the stroke. The red dot in both panels indicates the moment of the maximum velocity in the paddles speed. The red line in the bottom panel indicates the minimal distance between the paddle center at peak velocity (red dot) and the ball trajectory (blue line).

51

Procedure

Participants were placed in the middle of the tracking hall and were equipped with a HMD and a paddle. Participants were instructed to hit the ball that was played by the virtual agent as natural and accurate as possible. Participants were not instructed to hit the ball back to the virtual player. Before the experiment started participants completed a practice block which familiarized participants with the VE equipment and the task. In this practice block participants hit virtual static balls that were presented at eight different locations above the participant's end of the table. If a ball was hit a sound was played, the ball disappeared and the next ball appeared at a new location. After successfully hitting eight balls (with a radial error less than 10cm) the exercise was completed and the experiment started.

Each participant played 3 experimental blocks. Each experimental block consisted of the repeated presentation out of four different stroke animation sequences. Two stroke animation sequences showed a stroke with a valid initial posture cue (one was played to the left and the other to the right side of the participant). The other two stroke animation sequences showed an invalid initial posture cue (one was played to the left and the other to the right side of the participant). In each experimental block this four different stroke animation sequences were repeated for each of the twenty experimental conditions (=80 trials). The cells of the twenty experimental conditions were derived from the fully crossed design: 4 sources of visual information x 5 presentation durations. The order of the 80 trials in each block was randomized. Each trial started with a 400ms long beep sound to indicate the start of the trial. Afterwards one out of four stroke animations sequences was presented. In each stroke animation sequence the ball started flying from the participant's side of the table towards the virtual player's side of the table. Then the virtual player returned the ball back to the participant who attempted to hit the ball. A visual (text: "HIT" which was placed in the center of the screen for 1 second) and acoustic feedback (500ms sound) informed the participant when he/she hit the ball within a range of 50cm (distance between ball and paddle of the participant). There was no visual or acoustic feedback if the ball was missed. The feedback was purely motivational. Importantly, all trials were used for data analysis, whether the feedback was provided or not. In order to continue to the next trial participants were required to move back into the initial body posture (standing in the center with the right arm relaxed so that the paddle was positioned in parallel to the right thigh).

Results

Direction accuracy (DA) measured the percentage of trials in which participants executed the stroke on the same side of the table as the ball arrived. If participants randomly picked a side for executing a stroke DA should be equal to 50%. Likewise a DA of less than 50% indicates that participants executed more strokes to the opposite side (incorrect side) of the table than to the same side of the table where the ball arrived (correct side). We reasoned if participants use a given source of visual information to make a prediction about the location of the ball, DA scores should be significantly different from 50% in the associated experimental condition. In order to test when DA scores in the experimental conditions are significantly different from the 50% chance level we conducted t-tests on the average DA scores in the different experimental conditions. The 40 t-tests in the different experimental conditions (4 sources of visual information x 5 presentation durations x 2 posture cue validity) were conducted using Bonferroni adjusted alpha levels of 0.0025 per test (.05/40). The results of all t-tests are reported in Table 1. Figure 6 summarizes these results graphically by showing the average DA scores for each experimental condition separately. Figure 6a shows the average DA scores for the valid posture cue conditions. Figure 6b shows the average DA scores for the invalid posture cue conditions. The black dotted line in figure 6 shows the 50 % chance level which indicates that participants picked a side randomly. The error bars indicate the 95% Bonferroni corrected confidence intervals. An error bar that does not cross the line of the 50 % chance level indicates that DA in this experimental condition was significantly different from chance level. A significant effect was also indicated by a solid marker on the mean score.

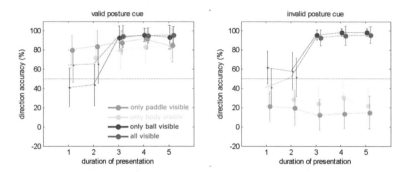

Figure 6: Overview of the average directional accuracy scores in the different experimental conditions. The left panel shows the average directional accuracy scores for the valid posture cue conditions. The right panel shows the average directional accuracy scores for the invalid posture cue condition. The black dotted line shows the 50 % chance level which indicates how participants should perform randomly. The error bars indicate the confidence intervals as derived from the t-tests. If an error bar did not cross the line of the 50 % chance level, this would indicate that directional accuracy in this experimental condition was significantly different from the 50 % chance level. A significant effect was also indicated by a solid marker on the mean score.

In the valid posture cue condition, we found DA to be significantly increased above the 50% level both when participants saw only the body or only the paddle of the virtual agent (see Figure 6a). Hence participants were able to choose the correct side of the table for striking the ball in these conditions. On the other hand in the invalid posture cue condition, we found that participants significantly more often chose the side of the table opposite to the side on which the ball actually arrived (see figure 6b). This suggests that participants relied on visual information about the virtual agent's body and paddle in order to choose the side on which they execute their strokes when no other visual information was available.

The results from the only paddle visible condition in Figure 6 are in line with our prediction that participants were able to use visual information about the virtual agent in order to infer the striking direction. In the valid postural cue was effective already for the shortest presentation duration in the only body visible condition as indicated by a DA that was significantly larger than 50%. That is, participants were able to infer the correct side of the table from the paddle information before the virtual agent actually executed any action (time point 1). Interestingly this pattern reversed in the invalid cue condition. Here participants exhibited a DA significantly lower than 50% for all presentation times in the only paddle visible condition. Hence participants always chose the incorrect side of the table as it was indicated by the postural cue. An interesting aspect is that participants did not change their prediction about which side the ball will arrive even after they have seen the virtual agent execute the stroke (time point 3 and larger). Hence the visual paddle information during action execution was not indicative about the side to which the virtual agent will play the ball. This finding further supports the idea that only the postural cue (which was in line with the opposite side of the table) but not the actual executed action (which was in line with the correct side of the table) guides the decision about where the virtual agent will play the ball. Overall the results of the only paddle visible conditions are in line with the idea that postural cues are used for predicting temporally distal events.

The results of the only body visible condition resemble those of the only paddle visible condition. The major difference is that the body produces a significant bias of decision about the playing direction from time point 2 on rather than from time point 1 in both the valid and invalid postural cue condition. Again since the prediction of the playing direction does not change after the virtual agent hit the ball, the overall results indicate that the 'posture' of body information is mainly driving participants' decision from early on about which side to execute the stroke.

In contrast, the only ball visible condition shows different results. In this condition participants did not prefer to execute their stroke to any particular side until time point 3 in both the valid and invalid cue condition. From this time point on, however participants always chose the correct side as indicated by DA values significantly above 50% in both the invalid and valid condition. Because time point 3 is the time point where participant actually saw the virtual agent hitting the ball, participants seem to use this event in order to infer the playing direction of the virtual agent. Overall since DA deviated significantly from chance level at a later point in time in the only ball visible condition (time point 3) than in the conditions containing visual information about the virtual agent (only paddle and ball visible condition) participants use the ball Information about the ball at a later time than visual information about the virtual agent. Hence these results are in line with our prediction that participants use visual information about another person at an earlier point of time than visual information about the ball.

Finally, the all visible condition is associated with DA values that are significantly different from the ball condition at time points 1 and 2 in the valid cue condition. However at time points 3, 4, and 5 the all visible condition is not significantly different from the only ball visible condition in the valid cue condition. The significant difference between the DA scores in the all visible condition and the DA scores in the only ball visible condition at time point 1 and 2 suggests that participants did not only rely on the ball information in the all visible condition during this time period. Because the all visible condition displays also visual information about the virtual agent's body and paddle, the just mentioned significant differences at time points at 1 and 2 indicate that participants' performance was also influenced by the presence of visual information about the agent. This observation is in line with the idea that participants use visual information about the other person for temporally distant events.

The pattern of the all visible condition in the invalid cue condition is similar to the observed pattern in the valid cue condition. The difference is that the all visible condition differs only significantly from the only ball visible condition at time point 1 suggesting that at this point of time participants' performance was affected by the additional information presented (virtual agents body and paddle) in the all visible condition. Hence, we also find evidence that participant's performance is affected early on by the virtual agent's visual information in the invalid cue condition.

Discussion

In this study we tested the hypothesis that different sources of information aid the interaction with a physical object within a social setting at different times. Specifically, we hypothesized that seeing the interaction object aids the prediction of temporally proximal events as it the case e.g. during online control of movements (Bootsma & Vanwieringen, 1990; Grierson, et al., 2009; McBeath, et al., 1995; McLeod & Dienes, 1993a; Sarlegna & Blouin, 2010) while seeing the interaction partner should aid the prediction of temporally distant events as it is the case e.g. for predicting future actions of a person (Vesper, et al., 2010). We therefore predicted that different sources of visual information (e.g. about the interaction object and about the interaction partner) should be important at different time phases of the social interaction. We further sought to find evidence for the idea that the information about the other person is used for prediction of the outcome of an action.

We tested this hypothesis in a task in which participants stroked table tennis balls which were played by a virtual agent within an immersive virtual environment. We manipulated the available sources of visual information and the presentation time for each source of visual information separately to assess when a particular source of visual information became important. We reasoned that if a source of visual information is important for action prediction then participants' performance should show a significant change at that time point. Further, we manipulated the validity of the initial posture cue of the virtual agent in order to test whether visual information about the virtual agent is used in order to predict where the virtual agent will stroke the ball. If information about the other person is used for predicting the outcome of the other persons actions, we expected that participants will predict always the same outcome (diagonal strokes) for the action. Consequently we expected that participants will predict for the ball to land on the incorrect side of the table in the invalid posture cue condition but on the correct side in the valid posture cue condition.

The results showed that visual information about the virtual agent (paddle and body) affects participants' performance to predict the side at which the ball will arrive. Indeed participants predicted for the ball to land on the incorrect side of the table in the invalid posture cue condition and on the correct side of the table in the valid postural cue condition. Interestingly this prediction pattern emerged even before the virtual agent hit the ball and did not change even after participants had seen the other agent executing the stroke. Hence participants seem to take the initial posture as a cue as to where another person will hit the ball.

Furthermore we found that visual information about the interaction partner is used at an earlier time than visual information about the ball. As outlined above our results suggest that participants used visual information about the other person to predict the side for the ball to arrive. Participants seem to use this cue right from the beginning, i.e. even before the virtual agent hit the ball. A different pattern was found when participants only saw the ball. Here, we found that seeing only the ball improved directional accuracy significantly after the virtual agent hit the ball but did not improve directional accuracy in the time phases before the virtual player hit the ball. Hence visual information about the ball influenced participants playing performance at a later point in time than seeing visual information about the virtual agent.

Interestingly when all information was visible directional accuracy was significantly improved in the valid posture cue condition and significantly impaired in the invalid posture cue condition compared to the only ball visible condition before the virtual agent hit the ball. This suggests that before the virtual agent hit the ball, visual information about the virtual agent's body and paddle was used in order to predict the fate of the stroke. Hence even if all visual is available participants relied on visual information about the interaction partner in order to predict temporally distant events (e.g. before the interaction partner performs an action on the object).

In sum, our results provided evidence that visual information about an interaction partner was used in order to predict temporally distant events. This is in line with the hypothesis that different sources of visual information operate on different predictive time scales.

Another interesting aspect about the findings reported in this study is that non-experts also seem to be able to predict the outcome of an action based on visual information about an interaction partner. Previous research often assumed that mainly experts are able to use visual information about another person to improve their performance in various tasks. For example, seeing the opponents body or interaction tool (i.e. the paddle or racket) improved ball prediction performance in tennis (Huys, et al., 2009; Mann, et al., 2010), squash (Abernethy, 1990), soccer (Savelsbergh, et al., 2002), basketball (Aglioti, et al., 2008), and volleyball (Urgesi, et al., 2011). The ability of naïve participants to use visual information about an interaction partner in order to predict temporally distant events should be beneficial to humans in everyday joint action tasks (Adolphs, 1999, 2003; Sebanz, et al., 2006).

The findings that both the paddle and body of the virtual agent was able to bias participants prediction about the arrival side of the ball suggests that both sources of visual information (paddle and body) were used in order to predict the action at the disposal of the virtual player and not in order to make physical predictions (e.g. determining the angle of reflection form the angle of inclination when the ball hits the racket). This question has not been tested before and provides important evidence about the role of visual information in social interactions. Our finding suggests that humans interacting with an object within a social setting (e.g. in a sports game) monitor and predict others actions in order to reduce task variability of temporally distant events.

Finally it is important to note that our experimental design allows for a higher external validity of the findings. Specifically, we employed a novel experimental paradigm which mimics a real life setting of social interaction and allows for a more natural interaction behavior. At the same time we had good control over the visual and temporal parameters required to test our hypothesis. In contrast to previous research which often investigated the processing of social stimuli using artificial tasks and restricted the way participants interact, we achieve a higher external validity of our findings with this novel approach.

Appendix

		Valid Posture Cue			Invalid Posture Cue		
Condition	Duration	Mean	Stdv	p	Mean	Stdv	p
only ball visible	1	41.27 %	26.67 %	0.149	61.90 %	23.05 %	0.0282
	2	43.65 %	28.61 %	0.321	57.93 %	26.15 %	0.1796
	3	92.86 %	16.30 %	***	96.03 %	7.27 %	0.0000
	4	96.03 %	8.98 %	***	98.41 %	5.01 %	0.0000
	5	93.65 %	16.22 %	***	98.41 %	5.01 %	0.0000
only body visible	1	61.90 %	19.82 %	0.012	34.92 %	21.01 %	0.0037
	2	72.22 %	23.17 %	***	28.57 %	18.36 %	***
	3	80.16 %	26.15 %	***	24.60 %	21.48 %	***
	4	83.33 %	22.36 %	***	30.95 %	21.91 %	***
	5	84.13 %	17.85 %	***	22.22 %	24.34 %	***
only paddle visible	1	80.16 %	20.83 %	***	21.42 %	21.17 %	***
	2	84.13 %	22.03 %	***	19.84 %	24.50 %	***
	3	88.10 %	16.79 %	***	12.69 %	21.01 %	***
	4	92.07 %	14.55 %	***	13.49 %	19.45 %	***
	5	84.92 %	22.92 %	***	15.07 %	22.30 %	***
all visible	1	64.29 %	23.74 %	0.012	41.26 %	25.61 %	0.1340
	2	65.87 %	26.60 %	0.012	51.58 %	26.82 %	0.7890
	3	94.44 %	15.21 %	***	92.85 %	11.26 %	***
	4	95.24 %	10.73 %	***	95.23 %	13.06 %	***
	5	96.03 %	11.67 %	***	96.03 %	11.67 %	***

Table 1: Overview of the statistics conducted on the directional accuracy scores.

Presentation Duration	Mean (only ball)	Mean (all visible)	t	Stdv	p
1	-0.1746	0.2857	-5.2597	0.4011	0.0000
2	-0.1270	0.3175	-3.4518	0.5900	0.0025
3	0.8571	0.8889	-0.5680	0.2561	0.5764
4	0.9206	0.9048	0.4385	0.1659	0.6657
5	0.8730	0.9206	-0.6794	0.3212	0.5047

Table 2: results of the paired t-tests conducted on the directional accuracy scores in the valid posture cue conditions.

Presentation Duration	Mean (only ball)	Mean (all visible)	t	STDV	p
1	0.2381	-0.1746	3.7489	0.5045	0.0013
2	0.1587	0.0317	0.8186	0.7108	0.4226
3	0.9206	0.8571	1.0000	0.2910	0.3293
4	0.9683	0,9048	1.2843	0.2265	0.2137
5	0.9683	0.9206	1.1421	0.1911	0.2669

Table 3: results of the paired t-tests conducted on the directional accuracy scores in the invalid posture cue conditions.

Presentation Duration	Mean (only ball)	Mean (all visible)	t	Stdv	P
1	0.7117	0.6297	3.3351	0.1126	0.0033
2	0.7291	0.6168	3.7237	0.1382	0.0013
3	0.4511	0.4346	0.8456	0.0895	0.4078
4	0.4277	0.4088	1.0049	0.0862	0.3269
5	0.4207	0.4066	0.8175	0.0794	0.4233

Table 4: results of the paired t-tests conducted on the spatial error scores in the valid posture cue conditions.

Presentation Duration	Mean (only ball)	Mean (all visible)	t	Stdv	p
1	0.5040	0.6234	-4.1323	0.1324	0.0005
2	0.5251	0.5579	-1.0576	0.1419	0.3029
3	0.3464	0.3626	-1.3214	0.0562	0.2013
4	0.3545	0.3573	-0.1740	0.0749	0.8636
5	0.3429	0.3371	0.3677	0.0727	0.7169

Table 5: results of the paired t-tests conducted on the spatial error scores in the invalid posture cue conditions.

Chapter 3: The Effect of Social Context on the Use of Visual Information

The current chapter was published:

Streuber, S., Knoblich, G., Sebanz, N., Bülthoff, H. H., & de la Rosa, S. (2011). The effect of social context on the use of visual information. *Experimental brain research*, *214*(2), 273-284.

Introduction

Humans are social beings and their interaction often requires the concerted coordination of actions in time and space to accomplish their goals (Sebanz, et al., 2006), for example when two people play table tennis. The correspondence between an individual's goals and the interaction partner's goals defines the social context (Manstead & Hewstone, 1996). If the goals of the interaction partners are in positive correspondence, for example when the goals are complementary or the same, the interaction partners cooperate. In contrast if the interaction partners' goals are in negative correspondence, the attainment of one person's goal results in the failure to achieve the other person's goal. In this case the interaction partners compete. The investigation of the effects of social context (i.e. of competition and cooperation) on an individual's behavior has a long standing history in social psychology (Triplett, 1898). More recently researchers have started to investigate the cognitive and neural processes involved in cooperative and competitive behavior during human interaction.

This research has shown that cooperation and competition are associated with different cortical activity as measured by fMRI (de Bruijn, de Lange, von Cramon, & Ullsperger, 2009; Decety, Jackson, Sommerville, Chaminade, & Meltzoff, 2004) and differences in behavior (Becchio, Sartori, Bulgheroni, & Castiello, 2008; Georgiou, et al., 2007; Ruys & Aarts, 2010). Specifically some of the latter studies suggest that action coordination in cooperative and competitive settings involves distinct motor planning mechanisms.

Georgiou et al. (2007) found that kinematic trajectories of the very same action are modulated by social context. Specifically they analyzed the kinematics of participants' reach-to-grasp movements towards a wooden block with different action goals. In the critical conditions participants either built a tower of blocks together with a co-actor in a cooperative fashion or they competed with a co-actor to place a block in the middle of the table first in order to build a tower. The kinematic patterns of the reach to grasp movement differed significantly from each other depending on whether the action goal was cooperative or competitive. Specifically, kinematic patterns of the two interaction partners were significantly correlated in the cooperative condition but not in the competitive condition. The authors suggested that the social context influences the social intentions which in turn affects motor planning and consequently results in different kinematic patterns during competitive and cooperative behavior.

Indeed, in a more recent study Becchio et al. (2008) found evidence that intentions alter kinematic patterns. In the critical conditions one participant was seated opposite to a confederate of the experimenter (a trained actor) at a table with two blocks in between them. The participant and the actor had to reach and grasp one block (reach-to-grasp phase) and then they stacked the objects on top of each other to build a tower (tower-building-phase). In the competitive condition participant and actor competed for placing the bottom block of the tower. In the cooperative condition the participant and actor were assigned roles as to who should build the bottom and the top part of the tower. To see whether intentions modulate the kinematic patterns of the participant, the actor showed incongruent behavior within a given social context on some trials (incongruent trials) prior to the actor's execution of an action. Specifically the actor showed a competitive attitude (in terms of her facial expression and body posture) in the cooperative condition and a cooperative attitude in the competitive condition. This change of attitude on these incongruent trials was confined to the reach-to-grasp phase of the actor's movement. Interestingly participants' kinematic patterns in the reach-to-grasp phase differed on incongruent and congruent trials suggesting that showing a different attitude and intention before the actual action influences the kinematic patterns (Becchio, et al., 2008). These results are in line with the idea that social context changes the intentions of the interaction partners which in turn affects motor planning and leads to different kinematic patterns.

Do changes of the social context only affect the way humans carry out motor actions or does it also affect the way they process visual information from the environment? If social context were to change the visual information that is important for a given task, this would provide further evidence for the idea that interacting with another person and acting alone rely on different psychological mechanisms (Becchio, et al., 2010; Knoblich & Sebanz, 2008). Furthermore identifying which visual information is most important in a given social context improves our understanding of the nature of the perceptual and cognitive processes that are at play in a particular social context.

How might social context affect the processing of task relevant visual information? One way in which social context might alter the processing of task relevant visual information is by changing the intentions of the interaction partners, which results in changes to their motor plans. If motor planning and visual information were closely linked one would expect that social context might also affect the way humans look at the environment in different social contexts. In line with this idea studies on eye gaze behavior during motor tasks suggest a close link between the eye gaze behavior and the particular task. The Investigation of gaze behavior during object interaction tasks reveals that participants look at task specific landmarks that are critical for the action control of the given task before the action is completed (R. S. Johansson, Westling, Backstrom, & Flanagan, 2001; Lee, Young, Reddish, Lough, & Clayton, 1983). For example, when participants were instructed to stack objects on top of each other, participants focused their gaze on the objects before they actually stacked one object on top of the other (Sailer, Flanagan, & Johansson, 2005). Johansson et al. (2001) suggested that the visual information at the gaze location is used for the motor planning.

This idea is supported by other research on online control of actions in object interaction tasks. These studies suggest that visual information is being used for the online control of action (Bootsma & Vanwieringen, 1990; Cressman, Cameron, Lam, I., & Chua, 2010; Grierson, et al., 2009; McBeath, et al., 1995; Mcleod & Dienes, 1993b; Sarlegna & Blouin, 2010). For example, baseball players adjust their catching behavior in an online fashion to disturbances of the baseball's flying trajectory (Fink, et al., 2009). These studies suggest a close link between visual information and motor planning. Taken together the link between social context and motor planning and the link between visual information and motor planning implies that social context also changes the way humans look at the environment.

65

We tested the hypothesis that social context modulates the use of visual information during social interactions by means of a table tennis task. Pairs of participants played a table tennis game in either a cooperative (Experiment 1) or competitive (Experiment 2) fashion. During the experiment we manipulated (for each player separately) the visibility of visual information about the players' rackets and body movements. We measured table tennis performance by means of the number of successful passes for each player separately. We reasoned that if a particular source of visual information is important for playing table tennis, rendering this source of visual information visible should positively affect the players' table tennis performance. We used this logic to assess the importance of different sources of visual information in different social contexts. If a particular source of visual information improves table tennis playing performance in one social setting (e.g. cooperation) but not in the other one (e.g. competition) it would indicate that the importance of this source of visual information was modulated by social context.

To this end we manipulated four sources of visual information in two different social contexts. We examined the effect of (1) the visibility of a player's own body, (2) the visibility of the other player's body, (3) the visibility of a player's own racket, and (4) the visibility of the other player's racket on the percentage of successful passes in cooperative (Experiment 1) and competitive (Experiment 2) table tennis play.

Experiment 1

The purpose of Experiment 1 was to investigate the importance of different sources of visual information during cooperative table tennis play. We manipulated the visibility of the racket and the body for each player separately. We decided to use point-light like stimuli for manipulating visual body information to ensure that the stimuli employed in the experiment highlight the dynamic aspect of an action and thereby the interactive component of the task. One class of stimuli that is well suited for this purpose are point-lights as they are deprived of figural cues and rich of motion cues (Johansson, 1973). Hence point light displays emphasis the dynamic aspect of an action and thereby the interactive component of our experiment.

Previous research showed that humans can infer action relevant information by observing the other person's racket and body. For example, the availability of visual information about the other player's racket improves the prediction of ball trajectories in tennis (Huys, et al., 2009; Mann, et al., 2010) and squash (Abernethy, 1990). Moreover participants fixate on the other player's racket when they predict a stroke (Ward, Williams, & Bennett, 2002). The improved prediction performance of the ball trajectory when seeing the other player's racket should lead to an increase of successful table tennis strokes in the current experiment.

Similarly, previous research suggests that participants can infer action intentions from observing the interaction partner's body. For example, humans are able to identify the intentions underlying observed body movements from point light stimuli (Barrett, et al., 2005; Runeson & Frykholm, 1983). Point lights are devoid of figural cues but preserve the essential movement kinematics of an action (Johansson, 1973). In previous research point-light stimuli were exclusively presented on video displays in order to demonstrate isolated observer's ability to detect the kind of actions performed (Dittrich, 1993; Vanrie & Verfaillie, 2004) and also the actor's expectations (Runeson & Frykholm, 1983) and intentions (Grezes, et al., 2004a). Knowing the intentions of the other player might facilitate performance because observers can predict what the other person is going to do next. For example, goal keepers can better predict the fate of a penalty kick when observing the body of the penalty kicker prior to ball contact (Savelsbergh, et al., 2002). Also basketball players can better predict the fate of a basketball shot when observing the body of the shooter (Aglioti, et al., 2008) before the ball is released from the hand. These results suggest that visual information about the other player's body enhances action prediction, which in turn should also improve the number of successful passes in a joint table tennis task.

However, most of these experiments were conducted under conditions that more closely resemble competitive than cooperative settings. One important difference between cooperative and competitive conditions is that the goals of the interaction partners are in line and therefore known to each other (Van Avermaet, 1996). Hence predicting goals should not be necessary in cooperative play. We therefore hypothesized that the visibility of the other player's racket and possibly the visibility of the other player's body should not affect cooperative table tennis performance.

Visual information about one's own arm is important for the online control of arm movements. For example, the visibility of one's own arm leads to improved reaching accuracy (Bard, Hay, & Fleury, 1985; Proteau, Boivin, Linossier, & Abahnini, 2000; Spijkers & Spellerberg, 1995) and faster adjustments of incorrect arm movements (Reichenbach, et al., 2009). In light of this we hypothesized that players' performance will benefit from the visibility of their own body due to a better online control of arm movements. We are not aware of any research examining the effect of the visibility of one's own racket on playing performance in racket sports. Hence the results of Experiment 1 will also help to shed light onto the effects of seeing one's own racket on playing performance.

Methods

Participants

28 right-handed participants were tested (mean age: 29.61; SD: 5.6). Data of one pair was lost due to a technical error. The data analysis was carried out on the data of the remaining 13 pairs (3 male pairs, 2 female pairs, and 8 mixed pairs). All participants had normal or corrected-to-normal vision. Participants were recruited from the Max Planck Institute Subject Database and were naive with respect to the purpose of the study. This research was performed in accordance with the ethical standards specified by the 1964 Declaration of Helsinki. All participants gave their informed consent prior to the experiment and received 8 Euros per hour for their participation.

Stimulus and Apparatus

Participants played table tennis in a windowless darkened room of 4x5 m. A standard table tennis table (length: 2.74 m, width: 1.53 m; height: 0.76 m) was located in the center of the room. The four corners of the table were painted with fluorescent paint. The top edge of the table tennis net was also painted with fluorescent paint. Two sets of two table tennis rackets were used. One set had the rim painted with fluorescent paint and the other consisted of normal rackets without the paint. Furthermore, fluorescent body markers (compressed cotton balls with a diameter of 3 cm) were attached with Velcro to a headband and black sweaters that participants wore on top of their clothes. The markers were placed at the wrist, elbow, shoulder, upper sternum, and forehead on both the left and the right side of the body. Fluorescent tape (30 x 3 cm) was attached at 1.5 m height to each of the four walls to avoid participants colliding with the walls when playing in the dark. The stimuli as seen from a participant's view in the different conditions are shown in Figure 1.

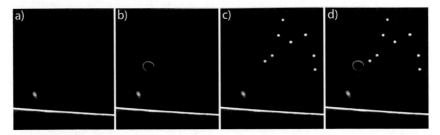

Figure 1: Images of experimental stimulus as seen from the perspective of one of the two participants. The ball, net, and table were visible in all viewing conditions. Panel a–c shows the three different viewing conditions in Experiment 1 and 2 from the perspective of one of the two participants. Eight experimental conditions were derived from a combination of these different viewing conditions for each participant of a pair. Panel d shows the experimental stimulus in the 'dark room' condition of the pilot study

A microphone was mounted under the middle of the table to record when the ball hit the table and to record participants' verbal responses (see below). The sound was recorded by means of custom written software on a computer. This computer also served for the manual recording of the hits and errors by the experimenter (see below).

The table tennis ball was also painted with fluorescent paint, which slightly changed its physical response properties (e.g. bouncing). However these changes did not affect the play as indicated by participants' reports. The same ball was used in all experimental conditions.

In order to validate the experimental environment a pilot study was performed in which 14 pairs of participants played cooperative table tennis in two different conditions. In the "light on" condition, participants played cooperative table tennis under normal light condition. In the "dark room" condition participants played cooperative table tennis with the self glowing markers attached to both participants and the rackets of both participants visible. Performance was measured as the percentage of successful passes out of 60 passes. The average performance score in the "light on" condition was slightly higher (mean performance: 93.83% of successful passes; SD: 4.82) than in the "dark room" condition (mean performance: 92.62% of successful passes; SD: 5.35). However, a paired t-test, $t(14)=1,32$, $p=0.210$, did not reveal a significant difference between the performance scores in both conditions. In sum this suggests that the body markers and the rackets provide all necessary information in order to reach a normal performance level (as in the lights on condition).

Design

The effect of visibility was investigated in eight experimental conditions. In all eight conditions the ball was always visible. In the 'Racket A' and 'Racket B' condition player A or player B was playing with a fluorescent racket, respectively. In the 'Racket A+B' condition the rackets of both participants were visible. In the 'No Racket' condition nothing else except for the ball was visible. In the 'Body A' and 'Body B' condition player A or player B was wearing the florescent body markers, respectively. The body markers resulted in the perception of a biological motion pattern of the player wearing the markers. In the 'Body A+B' condition both participants wore the fluorescent body markers. In the 'No Body' condition none of the two participants wore the fluorescent markers. From these eight conditions we derived three factors for the statistical analysis (factor 'own visibility' with levels visible/invisible; factor 'other player's visibility' with levels visible/invisible; and factor 'source of information' with levels racket/body) as outlined in

	Player A			Player B		
Conditions	Information	Own	Other	Information	Own	Other
Racket A	Racket	Visible	Invisible	Racket	Invisible	Visible
Racket B	Racket	Invisible	Visible	Racket	Visible	Invisible
Racket A+B	Racket	Visible	Visible	Racket	Visible	Visible
No Racket	Racket	Invisible	Invisible	Racket	Invisible	Invisible
Body A	Body	Visible	Invisible	Body	Invisible	Visible
Body B	Body	Invisible	Visible	Body	Visible	Invisible
Body A+B	Body	Visible	Visible	Body	Visible	Visible
No Body	Body	Invisible	Invisible	Body	Invisible	Invisible

Table 1: Overview of the eight experimental conditions coded with respect to the different viewing conditions as perceived by player A and player B.

Procedure

At the beginning of the experiment participants were informed about the following experimental procedure. Participants played table tennis according to standard table tennis rules with the additional instruction to play the ball back and forth as often as possible between them (cooperative play). Each pair of participants played each of the eight conditions three times for a total of 24 trials. The testing order of the experimental conditions was randomized across pairs of participants. Each trial consisted of 40 passes (playing the ball from player A to player B or vice versa). The experimenter turned off the lights before each trial and turned on the lights between trials to allow the fluorescent paint to recharge. The time between trials was used to inform participants about the specifics of the next trial and to equip each player with the appropriate items (fluorescent or non-fluorescent body markers and rackets) for the upcoming experimental condition. Then the experimenter switched off the lights and instructed one of the players to start with the serve after pressing a key on the keyboard, which resulted in playing the start sound of 2000 Hz. Participants only started playing after hearing the start sound. The experimenter pressed the space bar on the keyboard in

71

synchrony with the ball hitting the table to record the number of passes. The experimenter pressed either button A or B depending on who of the two players performed an error (the assignment of the labels A and B to participants did not change throughout the experiment and was only known to the experimenter). Each button press resulted in a distinct tone. The participant who committed an error was then loudly saying his/her name to have the name recorded by the microphone. After the ball was recovered the experimenter pressed the start button again (accompanied by a start sound) to indicate that the players could continue playing. The serve was alternated between the participants. The program counted the overall pass number in a trial and once the total number had been reached the program automatically played a stop sound to inform the participants about the end of the trial. Participants were not allowed to communicate verbally during the playing. The experiment lasted approximately 2 hours.

Results and Discussion

The factors for the statistical analysis coded which particular source of visual information was visible about the own or the other player's action: source of information (body vs. racket), own visibility (visible vs. invisible), and other player's visibility (visible vs. invisible). Importantly the dependent variable (percentage of successful passes) was measured for each player separately.

The results are shown in Figure 2a and 2b. Seeing the other player's racket and one's own body was associated with an improvement in performance. However, the visibility of the other player's body did not affect participants' performance. Surprisingly, seeing one's own racket was associated with a decrease in performance.

To investigate whether the observed effects bear statistical significance we tested the effect of source of information, own visibility, and the other player's visibility in a repeated measures ANCOVA. We aimed to control for the effect of the interaction partner's performance on one's own performance and used the interaction partner's percentage of successful passes as a covariate. The within subject factors of this ANCOVA were source of information (racket vs. body), own visibility (invisible vs. visible), and the visibility of the other player (invisible vs. visible).

72

The ANCOVA revealed a significant main effect of the other player's visibility (visible vs. invisible), $F(1,25) = 7.19$, $\eta^2_{partial}=0.222$, $p=0.013$, but no significant main effect of source of information (racket vs. body), $F(1,25)=0.33$, $\eta^2_{partial}=0.013$, $p=0.572$, and no significant main effect of own visibility (visible vs. invisible), $F(1,25)=0.07$, $\eta^2_{partial}=0.002$, $p=0.787$. There was also a significant interaction between source of information and own visibility, $F(1,25)=8.94$, $\eta^2_{partial}=0.260$, $p=0.005$, suggesting that seeing one's own racket and one's own body had different effects on playing performance. The interaction of source of information and the other player's visibility was also significant, $F(1,25)=5.93$, $\eta^2_{partial}=0.183$, $p=0.020$, indicating that seeing the other player's body and seeing the other player's racket differentially affected playing performances. The interaction between own visibility and the visibility of the other player was not significant, $F(1,25)=2.13$, $\eta^2_{partial}=0.076$, $p=0.155$. The three-way interaction between own visibility, the other player's visibility, and source of information was also non significant, $F(1,24)=0.17$, $\eta^2_{partial}=0.007$, $p=0.679$. There was no significant effect of the covariate, $F(24,1)=0.33$, $\eta^2_{partial}=0.953$, $p=0.569$.

Figure 2a shows the significant interaction between source of information and own visibility. Bars indicate the standard error from the mean derived from the appropriate error term of the interaction. Figure 2a shows that seeing one's own body has the opposite effect as seeing one's own racket. Paired t-tests were used in order to compare the effect of seeing one's own information on performance for each source of information separately. The percentage of successful passes was significantly higher when participants saw their own body compared to when they did not see their own body, $t(25)=2.697$, Cohen's $d=0.182$, $p=0.012$. On the other hand seeing one's own racket was associated with significantly worse playing performance than not seeing one's own racket, $t(25)=2.101$, Cohen's $d=0.142$, $p=0.046$.

The interaction between source of information and the visibility of the other player is shown in Figure 2b (bars indicate standard error). Paired t-tests were used in order to compare the effect of seeing the other player's information on performance for each source of information separately. The figure shows that seeing or not seeing the other person's body did not have an effect on the percentage of successful passes which is supported by a non-significant paired t-test, $t(25)=0.838$, Cohen's $d=0.053$, $p=0.410$. On the other hand seeing the other player's racket led to significantly better performance compared to when the racket was not visible, $t(25)=4.833$, Cohen's $d=0.306$, $p<0.001$.

In summary, we investigated the importance of different sources of visual information about one's own and the other player's actions on individual table tennis performance in cooperative table tennis. We found the largest positive change in performance when the racket of the interaction partner was rendered visible. The positive effect of seeing the other player's racket can be explained by the improved prediction accuracy of the ball trajectory in racket sports that is associated with seeing the other player's racket (Huys et al., 2009; Mann et al., 2010; Abernethy, 1990). A better prediction of the ball trajectory should lead to a better performance of hitting the ball, which in turn should result in better play. Performance also increased when one's own body was visible. Previous findings suggest that the visibility of one's own body contributes to improved online control of arm movements. The improved online control of the arm should result in increased contact with the ball thereby increasing playing performance.

Rendering the interaction partner's body visible did not change playing performance. Previous studies suggest that different sources of visual body information lead to different prediction accuracies of an action outcome (Savelsbergh, et al., 2002; Williams, Ward, Knowles, & Smeeton, 2002). There are several possible explanations as to why there was no improvement in performance when seeing the other player's body in Experiment 1. First, it is possible that participants did not anticipate the other player's action goals in Experiment 1. People who are cooperating often share action goals. Hence the goals of the interaction partner are typically known in cooperative tasks. For example, players might have known that the other person will return the ball in such a way that one is able to conveniently play back the ball in the current experiment. If players know about each other's action goals in cooperative table tennis play, no or very little prediction of goals should be necessary. As a result the visibility of the other body should have little effect. Finally, seeing one's own racket had a negative effect on playing performance. This finding is surprising since an obvious interpretation of this decrease is that seeing one's own racket is distracting.

To compare the use of visual information in cooperative and competitive context, Experiment 2 examined the importance of different sources of visual information in a competitive setting. Another set of participants played table tennis under the exact same conditions with the only exception that participants were instructed to play competitively.

Experiment 2

We expected that the importance of specific sources of visual information will be modulated by the context while other sources remain equally important in a cooperative and a competitive context. Specifically we expected that visual information about one's own body and the other player's racket will improve participants' performance for the same reasons as outlined in Experiment I. Therefore these sources of visual information should not be affected by the context modulation.

More importantly, we hypothesized that the visibility of the other player's body is crucial in competitive table tennis. Because action goals are not aligned in competitive settings (Van Avermaet, 1996), the action goals of the other player are unknown. A typical example is a penalty kick situation. Notice that the goals of the goal keeper (stopping the ball) and the kicker (scoring a goal) are not aligned. The goal keeper attempts to predict the corner to which the player will kick the ball to stop the ball while the player possibly attempts to predict the side to which the goal keeper will jump in order to score a goal. Hence action prediction should be much more important in competitive settings. In line with this idea effects of social intention are larger in competitive compared to cooperative situations (Decety, et al., 2004; Decety & Sommerville, 2003; Georgiou, et al., 2007). We hypothesized that participants should benefit from action prediction in competitive play and therefore seeing the other player's body should be important in competitive play.

As in Experiment 1, we assessed the effect of the visibility of one's own and the other player's racket and body on the number of successful passes.

Methods

The methods of Experiment 2 were identical to those of Experiment 1 with the following exceptions.

Participants

There were 28 participants (mean age: 28.18; sd: 3.32). All participants were right-handed and all had normal or corrected-to-normal vision. Participants were recruited from the Max Planck Institute Subject Database and were naive with respect to the purpose of the study. This research was performed in accordance with the ethical standards specified by the 1964 Declaration of Helsinki. All participants gave their informed consent prior to the experiment and received 8 Euros per hour for their participation.

Procedure

In Experiment 2 participants were instructed to play table tennis competitively by informing them that the participant with the least amount of errors would win the trial. There was no financial reward associated with winning a trial.

Results and Discussion

Experiment 2 set out to examine the effect of seeing one's own racket or body and seeing the other player's racket or body on the percentage of successful passes when table tennis is played competitively. The results of this experiment are shown in Figure 2c and 2d. Seeing one's own and the other player's body seems to improve performance. Furthermore seeing one's own and the other player's racket seems to have no impact on performance.

We examined the effect of source of information (racket vs. body), own visibility (visible vs. invisible) and other player's visibility (visible vs. invisible) on percentage of successful passes in a three factorial complete within-subject ANCOVA with the percentage of successful passes of the interaction partner as a covariate.

The ANCOVA revealed significant main effects of the visibility of the other player (visible vs. invisible), $F(1,27) = 10.57$, $\eta^2_{partial}=0.283$, $p=0.003$, and source of information (body vs. racket), $F(1,27) = 13.51$, $\eta^2_{partial}=0.307$, $p=0.001$, but no significant effect of own visibility (visible vs. invisible), $F(1,27)=1.46$, $\eta^2_{partial}=0.038$, $p=0.236$. The interaction between source of information and own visibility was significant, $F(1,27)=9.67$, $\eta^2_{partial}=0.263$, $p=0.004$. The interaction between source of information and the other player's visibility also turned out significant, $F(1,27)=5.78$, $\eta^2_{partial}=0.171$, $p=0.022$. The interaction between own visibility and the other player's visibility was not significant, $F(1,27)=0.20$, $\eta^2_{partial}=0.007$, $p=0.660$. The three-way interaction between the factors own visibility, the other player's visibility, and source of information was also non significant, $F(1,26)=0.62$, $\eta^2_{partial}=0.023$, $p=0.439$. There was also a significant effect of the covariate, $F(26,1)=15.45$, $\eta^2_{partial}=0.110$ $p=0.001$.

The significant interaction between source of information and own visibility is shown in Figure 2c. Paired t-tests were used to compare the effect of visibility of one's own information on performance for each source of information separately. Performance scores significantly improved when participants saw their own body compared to when their own body was invisible, $t(27)=3.816$, Cohen's $d=0.233$, $p<0.001$. One explanation of this result is that the visibility of one's own body lead to improved action coordination. We observed no significant change in performance when the visibility of one's own racket changed, $t(27)=0.685$, Cohen's $d=0.042$, $p=0.499$. Figure 2d shows a significant interaction between source of information and the visibility of the other player. Paired t-tests were used to compare the effect of visibility on performance for the other player's racket and the other player's body separately. The visibility of the other player's body led to an increase in the percentage of successful passes, $t(27)=4.585$, Cohen's $d=0.262$, $p<0.001$, while seeing the other player's racket did not lead to significant changes in performance, $t(27)=0.991$, Cohen's $d=0.057$, $p<0.331$. The result that visual information about the opponent improved playing performance supports our hypothesis that action prediction is critical in competitive play.

In a next step we directly compared Experiment 1 and 2 to determine the effect of social context on the importance of different sources of visual information.

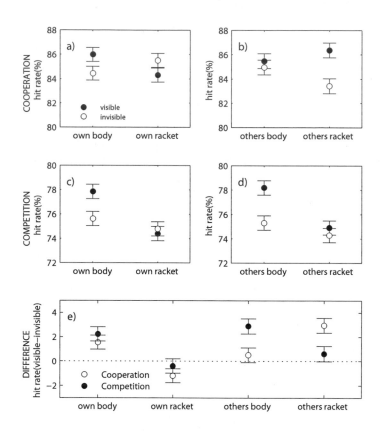

Figure 2: Mean performance scores of Experiment 1 (a, b) and Experiment 2 (c, d). The top two panels shows the effect of visibility of one's own information (a) and the other player's information (b) on mean performance scores in cooperative play. The middle panels show the effect of visibility of one's own information (c) and the visibility of the other player's information (d) on mean performance scores in competitive play. The bottom panel (e) shows the difference (visible–invisible) in performance for each source of information and for both contexts (cooperative vs. competitive). Error bars indicate the standard error from the mean.

Comparing cooperative and competitive play

We directly compared the results of Experiment 1 and 2 to estimate the effect of social context on the importance of different sources of visual information. A comparison of Figures 2a-2b and 2c-2d shows that cooperative play was associated with an overall higher performance than competitive play. Furthermore a comparison of the critical interactions in both experiments revealed that participants profited from seeing the other player's racket but not the other player's body in cooperative play (Figure 2b), whereas in competition participants profited from seeing the other player's body but not the other player's racket (2d).

To directly assess how social context modulates performance associated with the visibility of different sources of visual information we calculated the difference of the percentage of successful passes between visible and invisible conditions for each source of information and social context separately (Figure 2e). Positive differences indicate that the visibility of the information improved the percentage of successful passes while negative differences indicate a decrease in performance. Interestingly, the visibility of the participant's own racket and body led to similar performance changes in cooperative and competitive conditions. This suggests that social context did not change the importance of visual information about one's own movements. However, the pattern of results diverged regarding the visibility of visual information about the other player. Visibility of the other player's racket improved performance only in the cooperative condition whereas visibility of the other player's body improved performance only in the competitive condition. This indicates that social context modulates the importance of different sources of visual information.

To test whether this pattern bears statistical significance we compared the results of Experiment 1 and 2 in one overall analysis. Specifically we carried out an ANCOVA with context (competitive vs. cooperative) as a between subject factor and source of information, own visibility, and the other player's visibility as within subject factors. The performance of the other player was used as a covariate.

We found significant main effects of context (cooperative vs. competitive), $F(52,1)=24.65$, $\eta^2_{partial}=0.032$, $p<0.001$, the other player's visibility (visible vs. invisible),

$F_{(52,1)}=19.32$, $\eta^2_{partial}=0.272$, $p<0.001$, and source of information (tool vs. body), $F_{(52,1)}=8.29$, $\eta^2_{partial}=0.126$, $p=0.006$, but no significant effect of own visibility (visible vs. invisible), $F_{(52,1)}=1.81$, $\eta^2_{partial}=0.025$, $p=0.182$. The interaction between context and the other player's visibility was not significant, $F_{(52,1)}=0.01$, $\eta^2_{partial}<0.001$, $p=0.912$. The interaction between context and own visibility was also not significant, $F_{(52,1)}=0.40$, $\eta^2_{partial}=0.007$, $p=0.531$. Furthermore there was no significant interaction between context and source of information, $F_{(52,1)}=2.90$, $\eta^2_{partial}=0.046$, $p=0.094$. Also the interaction between the other player's visibility and source of information turned out to be non significant, $F_{(52,1)}=0.11$, $\eta^2_{partial}=0.002$, $p=0.744$. The interaction between the other player's visibility and own visibility turned out to be non significant as well, $F_{(52,1)}=2.92$, $\eta^2_{partial}=0.051$, $p=0.093$. On the other hand, the interaction between own visibility and source of information was significant, $F_{(52,1)}=2.92$, $\eta^2_{partial}=0.196$, $p=0.001$, indicating a performance difference associated with seeing one's own body and seeing one's own racket. The interaction between context, own visibility, and source of information, $F_{(54,1)}=0.24$, $\eta^2_{partial}=0.004$, $p=0.629$, however was not significant.

Importantly, the interaction between context, the other player's visibility, and source of information was significant, $F_{(52,1)}=6.00$, $\eta^2_{partial}=0.103$, $p=0.018$ suggesting that the social context had a differential effect on how the visibility of the other player's information (body vs. racket) affected table tennis performance. This result is in line with our hypothesis that social context modulates the importance of the visual information about the other person. Finally, there was no significant three-way-interaction between the other player's visibility, own visibility, and context, $F_{(52,1)}=1.24$, $\eta^2_{partial}=0.022$, $p=0.271$, no significant 3-way-interaction between the other player's visibility, own visibility, and source of information, $F_{(52,1)}<0.001$, $\eta^2_{partial}<0.001$, $p=1.000$, and no significant 4-way-interaction between context, the other player's visibility, own visibility, and source of information, $F_{(52,1)}=0.73$, $\eta^2_{partial}=0.014$, $p=0.396$. There was also a significant effect of the covariate, $F_{(51,1)}=6.85$, $\eta^2_{partial}=0.940$, $p=0.012$.

It could be that the differential effect of seeing the other player on table tennis performance in Experiment 1 and 2 was due to differences in playing speed rather than differences in social context. If players played faster in the competitive than in the cooperative condition they might have had less time to prepare their own strokes in the competitive condition. It is possible that participants might have looked for early cues about how the other player plays the ball by focusing on the other player's body cues. Indeed participants played significantly faster in the competitive (mean pass duration = 722 ms; SD=8.0 ms) than in the cooperative condition (mean pass duration = 923 ms; SD=13.0 ms), as revealed by an independent between samples t-test, $t(52)=6.89$, $p<0.001$. To see whether the modulation of the other player's information by social context can be explained by playing speed, we used the playing speed as measured by the pass duration as a covariate. The pass duration is the time between the moments when the player hits the ball to when the interaction partner hits the ball. We calculated the average pass duration for each trial and used this data as a covariate in the previous analysis. If the modulation of the import sources of visual information about the interaction partner was due to different playing speeds in different contexts we expect the interaction between context, visibility of the other player, and source of information to be no longer significant.

We ran the previous analysis, which compared Experiment 1 and 2, in exactly the same way with pass duration as an additional covariate. For sake of clarity we limit the report of this analysis to the critical interactions. The interaction between context, the other player's visibility, and source of information was significant, $F(54,1)=6.14$, $p=0.017$, again suggesting that social context modulated how the visibility of the other player's information affected performance. The interaction between context, own visibility, and source of information was not significant, $F(54,1)=0.26$, $p=0.601$. The significant three-way interaction between source of information, the other player's visibility, and context suggests that different playing speeds in the two social contexts cannot explain the differences in how the visibility of the other player's information affected table tennis performance.

In summary, the direct comparison of Experiment 1 and 2 shows that social context modulates the importance of the others player's visual information. This result cannot be explained by the faster playing in the competitive condition alone.

Discussion

In the current study we sought to examine how different sources of visual information affect table tennis performance in different social contexts. We therefore manipulated the visibility of one's own racket, one's own body, the racket of the other player and the body of the other player in a cooperative (Experiment 1) and competitive (Experiment 2) table tennis setting. The results showed that social context had a differential effect on table tennis performance depending on whether information about oneself or the other player was rendered visible. Manipulating the visibility of visual information about oneself had the same effect on table tennis performance in a competitive and in a cooperative social context. However, social context affected how information about the other player was used. Specifically, in the cooperative setting the most pronounced performance increases occurred when the other player's racket was rendered visible. In contrast, in the competitive condition rendering the other player's body visible was associated with the largest positive performance changes. This suggests that different sources of visual information are used in table tennis performances in competitive and cooperative context. Overall these results suggest that social context affects the importance of visual information about others.

Our results argue against the idea that the effects of social context on playing performance merely reflect the effect of the different playing speeds in cooperative and competitive play. When including playing speed as a covariate in the analysis, we found the same effects as in the analysis without playing speed as a covariate. We therefore deem it unlikely that playing speed is the sole mediator for the observed effect.

Our findings indicate that action prediction is more important in competitive than in cooperative play. As the goals of interaction partners align in cooperative play interaction partners can easily predict each other's actions. In contrast if the goals are not aligned (as in competitive play), action goals need to be inferred. Thereby, visual body information might serve as an important source of information. In line with this suggestion it has been shown that humans are able to infer intentions from point light stimuli (Barrett, et al., 2005; Runeson & Frykholm, 1983) and intentions influence behavior more strongly in competitive than cooperative settings (Georgiou, et al., 2007).

A possible alternative explanation for the finding that body information was more important in Experiment 2 than in Experiment 1 is that players in Experiment 2 may have been more experienced. Previous findings have shown that expert players focus on different parts of their partner's body during the anticipation of an action compared to novices and that they are better at predicting action outcomes (Savelsbergh, et al., 2002; Williams, et al., 2002). We therefore compared the experience of players between the two experimental groups. We measured table tennis experience of participants in terms of the amount of time participants played table tennis in the past year. The two samples t-test revealed no significant differences in table tennis experience between the two groups, $t(40)=1.73$, $p=0.091$. Hence motor expertise alone cannot explain our findings.

Because action goals are known in cooperative play, players might focus on different aspects of the task to improve the attainment of their action goals. For example players might have focused on the exact prediction of the ball trajectory in the cooperative condition to ensure that they play the ball in a way that it is optimal for the other player. Because the orientation of the racket and the angle of incidence is important to calculate the angle of reflection seeing the other player's racket might have become important.

Furthermore, we found that participant's performance improved when one's own body was visible but not when one's own racket was visible, independently from the social context. In line with previous research visual information about one's own body might have contributed to improved online control of arm movements which resulted in increased playing performance. The absence of an effect of seeing one's own racket might have been to due to the orientation of the racket in the participant's hand. As mentioned above seeing the racket might be important for predicting the ball trajectory (angle of reflection equals angle of incidence). However, the predictability strongly depends on the viewing angle. Participants saw their own racket in the periphery only and the viewing angle might be very inconvenient to make physical predictions about the ball trajectory. Therefore, participants might not have been able to use visual information about their own racket in order to facilitate their playing performance.

In order to investigate the effect of social context on the use of visual information we employed a novel experimental paradigm which takes into account the perceptual and motor interdependencies between two individuals performing a social interaction task. In the past, researchers often investigated the processing of social stimuli in isolated individuals. For instance, researchers in sport sciences investigated the importance of perceiving visual information about opponent player's actions using psychophysical methods. In most of these studies participants were asked to judge the fate of an action (e.g. a tennis shot) which was previously video recorded and finally displayed on a computer screen (Abernethy, 1990; Aglioti, et al., 2008; Huys, et al., 2009). The authors used spatial and temporal occlusion to test the effect of visibility on participant's prediction accuracy. The advantage of using psychophysical methods to examine human's ability to pick up task relevant information is the high degree of control and thus statistical power. However, it is not clear in how far this paradigm accounts for real life interactions in which two or more individuals influence each other's actions and are set in a common social context. The investigation of social interaction behavior under real-life conditions allowed for a more realistic assessment of the critical sources of visual information. Our findings point to a novel factor that influences the use of visual information. So far studies have shown that novices and experts focus on different sources of visual information suggesting that motor expertise is a critical factor in the use of visual information (Aglioti, et al., 2008; Calvo-Merino, et al., 2005; Casile & Giese, 2006; Keller, et al., 2007). Here we demonstrated that social context also modulates the importance of different sources of visual information.

Acknowledgements

We would like to thank Megan McCool for fruitful discussions and proof reading of the manuscript and Magdalena Leshtanska for part of the data recording. We gratefully acknowledge the support of the Max Planck Society and the WCU (World Class University) program through the National Research Foundation of Korea funded by the Ministry of Education, Science and Technology (R31-2008-000-10008-0). Natalie Sebanz acknowledges the support from a European Young Investigator Award by the European Science Foundation. The contribution of Stephan de la Rosa was funded by the EU Project TANGO (ICT-2009-C 249858).

General Discussion

The general aim of this thesis was to examine the use of visual information in joint action under close-to-natural conditions. This aim was accomplished through a number of studies using more natural experimental settings than previous studies. In three studies I examined the contribution of different sources of visual information (Study 1), the functional role of different sources of visual information (Study 2), and the effect of social context on the use of visual information (Study 3) in a joint action table tennis game. Following I will present the main findings of these studies and discuss them in relation to current literature. I will also outline some ideas about future research possibilities.

Studying Joint Action In the Virtual Environment

The aim of this thesis was to examine the role of visual information in joint action under close-to-natural conditions. This aim requires methodological approaches that enable natural interaction. Standard approaches that have been used in joint action research often relied on standard laboratory settings (e.g. monitor screen and button box for responses). The advantage of these approaches is that it enables a high degree of experimental control (Abernethy, 1988, 1990; Abernethy & Russell, 1987; Abernethy & Zawi, 2007; Abernethy, et al., 2008; Aglioti, et al., 2008; Farrow & Abernethy, 2003; Houlston & Lowes, 1993; Huys, et al., 2009; Isaacs & Finch, 1983; McMorris & Colenso, 1996; Penrose & Roach, 1995; Renshaw & Fairweather, 2000; Salmela & Fiorito, 1979; Urgesi, et al., 2011; Weissensteiner, et al., 2008). On the flipside standard laboratory settings do not allow for a natural interaction with an environment. To overcome previous shortcomings I used virtual environments (VE). VEs are increasingly frequently utilized in neural and behavioral sciences to investigate human behavior in complex environments (Bülthoff, 2009; Durgin & Li, 2010; Fink, et al., 2009; Loomis, Blascovich, & Beall, 1999; N. Vignais et al., 2009; Wann & Rushton, 1995; Warren, Kay, Zosh, Duchon, & Sahuc, 2001). The major advantage of this approach is that one achieves both a high degree of experimental control and natural interaction in complex environments in contrast to standard laboratory settings. For instance, previous studies

85

examined the participants' ability to predict the stroking direction by having participants press a left and right button on a button box in response to video presented on a computer screen. Here I used a head mounted display to present participants a stereoscopic image of a computer generated person (virtual player) from an egocentric perspective. Participant head and paddle movements were tracked in real time and directly translated into the VE. This setup allows both a natural interaction and a high degree of experimental control. In terms of natural interaction, the participant was required to hit virtual table tennis balls with a paddle in response to the virtual player. Hence, the setup of the current studies closely mimics the real world task of hitting a table tennis ball. At the same time the VE setup gives the opportunity to have a high degree of experimental control over a complex visual environment. For instance, in study 2 I was able to precisely manipulate visual information during table tennis play (e.g. hiding the ball when it was hit by the interaction partner). VEs are therefore a good research tool for advancing joint action research towards examining joint actions under more natural settings that allow a high degree of experimental control.

Several other advantages arise from having participants naturally interact with their environment. First of all, the natural linkage between action and perception that occurs in many everyday tasks is preserved under these conditions. This might result in the use of different response strategies on the participants' side. For example in standard laboratory settings participants might only make a prediction about the fate of the ball but under more natural settings participants might make this prediction for generating an adequate motor response. The altered purpose of the prediction might change participants' behavior. Overall the behavior in the VE should be more ecologically valid compared to standard laboratory settings. Secondly these more ecologically valid conditions opens the possibility of new behavioral measures that allow a more detailed analysis of human behavior. For instance, the VE allowed me to measure human behavior over the entire time course of the interaction (e.g. in terms of the 3D-movement kinematics) in contrast to e.g. a button press which measures mainly the action effect (Aglioti, et al., 2008).

Study 1 and Study 2 took advantage of these aspects. The VE was used to test the contribution of visual information on participant's behavior when performing table tennis strokes in response to the virtual player. This allowed gaining more insight onto the importance of different sources of visual information (Study 1) and the timing when these sources become important (Study 2). Visual information about the ball, the paddle and the body of the virtual player was selectively rendered invisible during play. The effect of the occlusions on participant's performance was measured in terms of task performance (radial error, directional accuracy) and also movement kinematics (stroke speed variability).

The main finding of Study 1 was that the condition in which only the ball was visible and the condition in which all sources of visual information was visible resulted in very similar striking performance of participants. This suggests that participants mainly rely on visual information about the ball. However when participants were forced to rely on visual information about the virtual player (i.e. when the visual information about the ball was turned off) participants were able to improve their striking accuracy significantly. Taken together these results suggest that participants prefer to rely on the visual information about the ball but have the ability to use visual information about the other player to improve their performance. Moreover the analysis of stroke speed profiles revealed that participants' striking speed is less variable when visual information about the other person is visible. Interestingly participants showed a decreased level of stroke speed variability when the virtual player was visible (body or paddle). This finding supports the idea that visual information about the other person guides one's own behavior. Overall, Study 1 indicates that visual information about the other player is important in a table tennis task even for untrained participants.

Study 2 examined the temporal dimension with regards to when different sources of visual information become important during an interaction. Based on theoretical considerations, we reasoned that ball information should become important at a later point in time than visual information about the other player. The results showed that visual information about the other player (body and paddle) biased or primed participants' prediction about the stroke direction. Specifically, participants' prediction about the virtual player's playing direction were in line with the virtual player's body posture regardless of whether the body posture indicated the correct or incorrect playing direction. This effect occurred before the virtual player actually hit the ball and maintained even after the participant hit the ball. In contrast participants were able to make accurate predictions about the virtual player's stroking direction after the virtual player had hit the ball. Hence, different sources of visual information were used at different temporal phases during the interaction.

The results of study 2 imply that participants use visual information about the virtual player to take a first guess about the playing direction. Such a strategy would be useful to achieve a more efficient motor planning, e.g. moving the center of body mass appropriately. Importantly this guess about the playing direction is not an accurate one. If the virtual player's posture indicated the incorrect playing direction, participants' predicted the incorrect direction. In this condition participants were unable to predict the correct playing direction even after they had seen the virtual player stroke. A comparison of the 'all visible' with the 'only ball visible' condition is interesting as it might allow some speculation about how different sources of visual information might be used. The predictions of the playing direction are significantly different in the 'all visible' condition from the 'only ball visible' condition before the virtual player hits the ball suggesting that participants relied on visual information about the virtual player (paddle and body) in the all visible condition during this time. After the virtual player made contact with the ball, the 'all visible' condition and the 'only ball visible' condition are very similar. Based on these results one might speculate that participants used visual information about the other player to take a first guess about the playing direction (hence the significant difference between 'the only ball' visible and 'all visible' condition). This first guess is then refined by more precise information about the ball's flying trajectory (hence the very similar performance between only ball and all visible condition after the virtual player made contact with the ball).

The results of Study 2 are overall in line with that of Study 1. First of all, Study 2 suggests that visual information about the ball is mainly used to make an accurate prediction about the ball. According to this result one would expect that stroking accuracy should be highest for the ball, which is what we found in Study 1. On the other hand this is not the entire story. Study 2 shows that participants use other sources of visual information, in particular the visual information about the other player in order to predict the hitting direction. The finding in study 1 that participants can use visual information about the other person's body to reduce their radial error if they are forced to is in line with this idea. Hence the results of both studies (1 & 2) support the idea that naïve participants are able to use visual information about the other person.

88

Based on these findings one might speculate that visual information about the other person guides one's behavior in terms of indicating the overall playing direction and preparing one's action for this event. This hypothesis might also be able to explain the increased stroke speed variability in Study 1 before the virtual player hits the ball in the condition where only the ball was visible. The lack of this guiding visual information might lead to the increase in stroke speed variability in the ball only condition of study 1,

Finally the assumption derived from study 2 that the visual information about the other player is used to make an overall guess about the playing direction before this guess is further refined by the ball information might be able to explain some of the differential effects between the two dependent variables in study 1. To reiterate, I found a strong effect of the ball on the radial error measure while the movement kinematics measure indicated an effect for the visual information about the other player. It is important to note that the radial error measures the position at the very end of an action (i.e. when the participant hit the ball). By the above hypothesis, the visual information about the ball should be most important shortly before the end of the stroke action and hence the radial error should be sensitive to the ball information. Because the visual information about the other player is important at an earlier point of time of the stroke action, radial error should be only slightly dependent on visual information about the other player. In contrast movement kinematics is a continuous measure over time and therefore can measure effects induced by the presence of certain visual information over the entire time course of an action (even at early times of an action). In this sense movement kinematics is a much more sensitive measure for an effect of visual information about the other player.

In sum using a novel experimental setting (the VE) allowed me to assess the importance of sources of visual information under more natural conditions. Maintaining a natural linkage between action and perception should result in findings that are more ecologically valid than in previous studies. Using the VE allowed me to uncover effects that have not been reported previously. I found that naïve participants are able to use visual information about the interaction partner and suggest that this information is used to make general predictions (i.e. the overall playing direction) about another person's actions.

Studying Joint Action in a "Closed Loop"

The VE allowed for a natural coupling of action and perception but not for reciprocal joint actions. In natural joint actions both interaction partners' actions are mutually coupled (closed loop). In the VE the participants responded to the virtual player's actions, but the virtual players actions were not influenced by the behavior of the participant (open loop). Hence, studying the role of visual information in the "closed loop" should further increase the naturalness of the interaction (Becchio, et al., 2010; U. Frith & Frith, 2010; Schilbach, et al., 2012; Sebanz, et al., 2006). In order to investigate joint action in the 'closed loop' I implemented a novel experimental setting that takes into account the social context of joint action. In the 'closed loop' environment pairs of participants could play table tennis while visual information about both players was manipulated selectively. This allowed for an assessment of the role of visual information under more realistic condition and within a social context.

Study 3 employed the 'closed loop' paradigm to investigate the importance of different sources of visual information in reciprocal joint actions within a social context. In two experiments, pairs of participants played table tennis either cooperatively or competitively. Visual information about both player's paddles and body kinematics was occluded selectively. Task performance was measured in terms of the number of successful passes (hit rate).

In study 3 the examination of the effect of visual information revealed that non-expert participants relied on visual information about interaction partner's actions in both social settings. This is different from previous studies which suggested that only highly trained experts are able to use visual information about the opponents movement kinematics in order to improve task performance (Abernethy, 1988; Abernethy & Russell, 1987; Aglioti, et al., 2008; Houlston & Lowes, 1993; Huys, et al., 2009; McMorris & Colenso, 1996; Penrose & Roach, 1995; Poulton, 1957; Van der Kamp, et al., 2008).

Moreover the social context modulated the importance of different sources of visual information about the interaction partner but not about the self. Seeing the own body always improved task performance irrespective of the social context. However, social context modulated the importance of the visual information about the other player. In the cooperative condition, seeing the other player's paddle had the largest influence on performance increase, whereas in the competitive condition, seeing the other player's body resulted in the largest performance increase. These results suggest that social context selectively modulates the use of visual information about others' actions. This latter finding provides evidence for the idea that it is necessary to include the social context in the analysis of joint action behavior. In the following paragraphs I will discuss the main findings in the context of current literature and ongoing research.

Social Cognition in Joint Action

The results of Study 1, 2, and 3 repeatedly showed that visual information about the interaction partner is critical for successful joint action under natural conditions. The finding that visual information about the interaction partner guides participant's behavior in an interceptive sports setting is novel. Previous research investigating visual anticipation in interceptive sports found that only highly trained expert players are able to pick up visual information about the opponent (e.g. about the paddle/racket or parts of the striking arm) but not novice players (Abernethy, 1988; Aglioti, et al., 2008). Many previous results are frequently interpreted as suggesting that only highly trained expert players are able to anticipate the actions at the disposal of the opponent while novices are only able to anticipate the ball flight trajectory (Abernethy, 1988; Abernethy & Russell, 1987; Aglioti, et al., 2008; Houlston & Lowes, 1993; McMorris & Colenso, 1996; Penrose & Roach, 1995; Poulton, 1957; Van der Kamp, et al., 2008). My results challenge this interpretation by providing empirical evidence that also novice table tennis player's behavior is guided by visual information about an interaction partner. For instance, Study 3 provided empirical evidence that novice table tennis player relied on visual information about the opponents body kinematics when performing a competitive table tennis task. One possible reason for the absence of an effect of visual information about an opponent on task performance might be that previous studies used standard laboratory tasks. For instance, previous research investigating visual anticipation in interceptive sports was predominantly examined using psychophysical occlusion paradigms (Abernethy, 1988, 1990; Abernethy & Russell, 1987; Abernethy & Zawi, 2007; Abernethy, et al., 2008; Aglioti, et al., 2008; Farrow & Abernethy, 2003; Houlston & Lowes, 1993; Huys, et al., 2009; Isaacs & Finch, 1983; McMorris & Colenso, 1996;

Penrose & Roach, 1995; Renshaw & Fairweather, 2000; Salmela & Fiorito, 1979; Urgesi, et al., 2011; Weissensteiner, et al., 2008). In this experimental environment participants had to press buttons in response to video clips presented on a computer screen. Constraining actions to button presses simplifies the experimental setup and provides a high degree of control and ability to replicate experiments. On the other hand this setup does not provide the linkage between action and perception observed in natural joint actions. However the linkage between action and perception might be critical for the use of visual information. For instance, when hitting table tennis balls players adjust the timing of their stroke to the approaching ball in an online fashion during stroke execution (Bootsma & Vanwieringen, 1990). The standard psychophysical setting does not provide a temporal coupling of visual information and participant's continuous behavior. However, a temporal coupling of action and perception might be critical for visual anticipation. Evidence for this comes from research examining tennis player's ability to predict the direction of tennis shots either verbally (uncoupled condition) or in situ by hitting a ball (coupled condition) (Farrow & Abernethy, 2003). The results revealed higher prediction accuracy in the coupled compared to the uncoupled condition. This illustrates that the absence of a natural action perception coupling partially prevents visual anticipation. Hence, the absence of an effect of visual anticipation in previous psychophysical studies might be due to reduced visual anticipation abilities by dissecting the natural linkage between action and perception.

The idea that action perception couplings affect visual anticipation is also supported by neurophysiological studies (Milner & Goodale, 1995; Milner & Goodale, 2008). Milner and Goodale (1995, 2008) argued that the programming of motor movements involves two functionally (and neuroanatomical) different visual processing streams. The 'ventral stream' processes visual information about the characteristics of objects and their relations for the purpose of perceptual identification and classification (vision for perception). On the other hand the 'dorsal stream' processes visual information for the online control of motor movements during action execution (perception of action). It has been suggested that sports performance requires the interacting contributions of the ventral system in order to perceive what action the situation affords and the dorsal system in order to visually guide one's actions (Van der Kamp, et al., 2008). Considering the two functional information processing streams with respect to the

92

experiment paradigm one might conclude that the uncoupled standard methods may have primarily engaged the 'ventral stream' whereas the VE engagement both the 'ventral stream' and the 'dorsal stream'. Hence, the absence of an effect in the previous studies using standard methods might be due to the unilateral engagement of the 'ventral system'.

The finding that visual information about the virtual player (body and paddle) was most critical for task performance at an early temporal phase of the interaction (before the virtual player hit the ball) indicates that participants were able to predict the action of the virtual player as suggested by theories of social cognition. However, relatively little is known about the underlying psychological mechanisms of human's ability to predict the fate of an action based on early visual information. One class of theories (simulation theories) suggest that humans predict actions of others by using one's own motor system to adopt the others perspective (Gallese & Goldman, 1998; Rizzolatti & Sinigaglia, 2010; Schilbach, et al., 2012; Sellars, 1956). These theories suggest that seeing the virtual player performing a goal directed action (e.g. performing a stroke) should allow for making predictions about the ongoing course of the interaction. However, the finding that the participants inferred the playing direction from the initial body posture (irrespective of the actual stroke) points to the possibility that participants partially relied on the situational probability of the occurrence of particular strokes on the basis of a static initial game configuration. Further research needs to be conducted to uncover the neural and psychological underpinnings of human's ability to predict the outcome of an action based on action observation.

Finally, the finding that visual anticipation of an interaction partner's actions is critical in joint action is in line with our daily-life experience. Anecdotal evidence shows that even novices can be tricked into believing that a stroke differs from the actual one that will be executed. For example, in a table tennis game even novices can be fooled into believing that a ball will be hit hard (by the other player displaying an appropriate preparatory hitting action) while in fact the ball will be played soft. Overall, visual anticipation is advantageous in many situations in every-day life. For instance, when shaking hands the responder usually anticipates the initiators arm movement instead of waiting until the hand is reached fully out. The results of this thesis provide empirical support for the intuition that visual anticipation is involved in every-day life joint actions.

Online Motor Control in Joint Action

Seeing the ball was the most relevant source of visual information for increasing task performance (Study 1 and Study 2). The finding that visual information about the ball was most critical for task performance at a later temporal phase of the interaction (after the ball was returned by the virtual player) is consistent with theories on online motor control. These theories suggest that visual information should be most critical shortly before the participant hits the ball. That idea that online motor control of behavior relies on visual ball information in various interceptive tasks is also supported by previous research (Bootsma & Vanwieringen, 1990; Craig, et al., 2006; Fink, et al., 2009; McBeath, et al., 1995; McLeod & Dienes, 1993a). However, it is important to point out that visual information about the ball was important shortly after the ball was returned by the virtual player and not as expected when the participant hit the ball. One reason of this finding might be that the technical configuration of the VE was suboptimal with respect to assessing online motor control mechanisms. This was due to the fact that the head mounted display's (nVisor SX60) field of view (vertical FOV of 34 and a horizontal FOV of 44) was limited. This limitation reduced participant's peripheral vision significantly (as compared to the real world). This technical limitation introduced the artifact that the ball was only visible within the field of view, but not when the participant actually hit the ball. In order to better investigate the processes underlying online motor control one must provide viewing conditions in which participants perceive the consequences of their own actions more similarly to the real world. One elegant solution to this problem might be to use walk in displays (CAVE technology) instead of head mounted displays. Walk in displays allow the user to be immersed within the VE without the necessity of wearing HMDs which limit the field of view. Using such an environment might allow a stronger emphasize on the question about how visual information about the ball translates into movement (e.g. by uncovering the underlying motor control laws). This would also allow testing if and how cognitive factors (e.g. the social context) affect online control in terms of the underlying control laws as suggested by (Georgiou, et al., 2007).

The Social Context

The results of Study 3 showed that social context modulated the importance of visual information about the other player. In the cooperative setting participants relied on visual information about the paddle of the interaction partner. However, in the competitive setting, participants relied on visual information about the body of the other player. This finding indicates that participants used visual information about the other player in both contexts. Interestingly, the finding that visual information about the interaction partner was important on top of seeing the ball is different from previous studies using "open loop" experiments. For instance, Study 1 & 2 (open loop) revealed that seeing the virtual player did not elicit an additional effect on top of seeing the ball. One possible explanation of the different findings is that hitting a ball in a 'closed loop' environment requires different mechanisms than hitting a ball in an 'open loop' environment. For instance, Study 1 & 2 (open loop) required participants to hit table tennis balls. Study 3 (closed loop), on the other hand, required participants to return table tennis balls towards the other player. The difference between both tasks is that the latter one involves the planning and execution of an action with respect to how this action affects the action at the disposal of the interaction partner. Visual information about the interaction partner might be most critical to perform such an action in reference to the other player. This might also explain the modulation of the importance of different sources of visual information by the social context. For instance, in the competitive condition, participants might require visual information about the interaction partner's body in order to decide where to place the ball so that the opponent cannot reach it. In the cooperative condition, however, participants might require visual information about the interaction partner's paddle in order to play the ball in a way that it can be easily hit by the interaction partner. In sum, the results of Study 3 suggest that the social context must be considered in joint action research. This requires a paradigm shift from studying the processing of social stimuli in isolated individuals towards investigation joint action in closed loop environments (Becchio, et al., 2010; U. Frith & Frith, 2010; Schilbach, et al., 2012; Sebanz, et al., 2006).

Further research should also address the question whether the effects elicited by the expertise manipulations in previous studies (Abernethy & Russell, 1987; Abernethy, et al., 2008; Aglioti, et al., 2008) and social context manipulations (in Study 3) share a common mechanism. For instance, previous studies found that only elite players (with motor experience), but no visual experts (without motor experience) or novices (Aglioti, et al., 2008) can read the body kinematics of an opponent player. This finding was interpreted to mean that motor expertise is required in order to read the body kinematics. However, the results of Study 3 suggest that also novice players can read the body kinematics of the opponent but that this ability affects performance only within a competitive social context. Therefore an alternative interpretation of Aglioti's results (Aglioti, et al., 2008) might be that elite players perceived the stimulus as being more competitive than the other groups (e.g. novices or visual experts). Hence, it would be interesting to test the effect of 'social context' and 'expertise' on the modulation of visual anticipation within a single study. For instance, it would be interesting to ask intermediate and professional players to judge the fate of basketball shots performed by team mates (which usually requires cooperative play) or by players from competing teams (which usually requires competitive play).

Afterthoughts

I used two novel experimental settings in order to investigate the importance of visual information in joint action under more natural conditions than previous studies. Specifically, I used a VE in order to provide a natural linkage between action and perception. Further I employed a 'closed loop' setting in order to investigate reciprocal joint actions in a social context. Using these experimental settings allowed me to uncover novel findings that have not been reported previously. Most importantly, I provided empirical evidence for the importance of visual information about the interaction partner in natural joint actions. I found that different sources of visual information were important at different times during the interaction suggesting the interplay between different mechanisms as described by theories on online motor control and theories on social cognition. Further, my results suggest that the social context affects the use of visual information. Hence, the social context must be considered in future research. This might require a paradigm shift from studying the processing of social stimuli in isolated individuals towards the investigation of natural joint action within a social context (Becchio, et al., 2010; U. Frith & Frith, 2010; Schilbach, et al., 2012; Sebanz, et al., 2006). Virtual Reality can provide a powerful research tool in order to foster this paradigm shift.

.

References

Abernethy, B. (1988). The Effects of Age and Expertise Upon Perceptual Skill Development in a Racquet Sport. *Research Quarterly for Exercise and Sport, 59*(3), 210-221.

Abernethy, B. (1990). Expertise, Visual-Search, and Information Pick-up in Squash. *Perception, 19*(1), 63-77.

Abernethy, B., & Russell, D. G. (1987). Expert Novice Differences in an Applied Selective Attention Task. *Journal of Sport Psychology, 9*(4), 326-345.

Abernethy, B., & Zawi, K. (2007). Pickup of essential kinematics underpins expert perception of movement patterns. *Journal of Motor Behavior, 39*(5), 353-367.

Abernethy, B., Zawi, K., & Jackson, R. C. (2008). Expertise and attunement to kinematic constraints. *Perception, 37*(6), 931-948.

Adolphs, R. (1999). Social cognition and the human brain. *Trends in Cognitive Sciences, 3*(12), 469-479.

Adolphs, R. (2003). Cognitive neuroscience of human social behaviour. *Nature Reviews Neuroscience, 4*(3), 165-178.

Aglioti, S. M., Cesari, P., Romani, M., & Urgesi, C. (2008). Action anticipation and motor resonance in elite basketball players. *Nature Neuroscience, 11*(9), 1109-1116.

Bard, C., Hay, L., & Fleury, M. (1985). Role of Peripheral-Vision in the Directional Control of Rapid Aiming Movements. *Canadian Journal of Psychology-Revue Canadienne De Psychologie, 39*(1), 151-161.

Barrett, H. C., Todd, P. M., Miller, G. F., & Blythe, P. W. (2005). Accurate judgments of intention from motion cues alone: A cross-cultural study. *Evolution and Human Behavior, 26*(4), 313-331.

Becchio, C., Sartori, L., Bulgheroni, M., & Castiello, U. (2008). Both your intention and mine are reflected in the kinematics of my reach-to-grasp movement. *Cognition, 106*(2), 894-912.

Becchio, C., Sartori, L., & Castiello, U. (2010). Toward You: The Social Side of Actions. *Current Directions in Psychological Science, 19*(3), 183-188.

Bideau, B., Multon, F., Kulpa, R., Fradet, L., Arnaldi, B., & Delamarche, P. (2004). Using virtual reality to analyze links between handball thrower kinematics and goalkeeper's reactions. *Neuroscience Letters, 372*(1-2), 119-122.

Bootsma, R. J., & Vanwieringen, P. C. W. (1990). Timing an Attacking Forehand Drive in Table Tennis. *Journal of Experimental Psychology-Human Perception and Performance, 16*(1), 21-29.

Bülthoff, H. H. (2009). Multisensory integration for perception and action in virtual environments. *Perception, 38*, 2-2.

Calvo-Merino, B., Glaser, D. E., Grezes, J., Passingham, R. E., & Haggard, P. (2005). Action observation and acquired motor skills: an FMRI study with expert dancers. *Cereb Cortex, 15*(8), 1243-1249.

Casile, A., & Giese, M. A. (2006). Nonvisual motor training influences biological motion perception. *Curr Biol, 16*(1), 69-74.

Chapman, S. (1968). Catching a Baseball. *American Journal of Physics, 36*(10), 868

Churchland, P. M. (1988). Matter and consciousness : a contemporary introduction to the philosophy of mind (Rev. ed. ed.). Cambridge, Mass. ; London: MIT.

Craig, C. M., Berton, E., Rao, G., Fernandez, L., & Bootsma, R. J. (2006). Judging where a ball will go: The case of curved free kicks in football. *Naturwissenschaften, 93*(2), 97-101.

Cressman, E., Cameron, B., Lam, M., I., F., & Chua, R. (2010). Movement duration does not affect automatic online control. *Human Movement Science*.

Davids, K. (2002). Interceptive actions in sport : information and movement. London ; New York: Routledge.

de Bruijn, E. R. A., de Lange, F. P., von Cramon, D. Y., & Ullsperger, M. (2009). When Errors Are Rewarding. *Journal of Neuroscience, 29*(39), 12183-12186.

Decety, J., Jackson, P. L., Sommerville, J. A., Chaminade, T., & Meltzoff, A. N. (2004). The neural bases of cooperation and competition: an fMRI investigation. *Neuroimage, 23*(2), 744-751.

Decety, J., & Sommerville, J. A. (2003). Shared representations between self and other: a social cognitive neuroscience view. *Trends in Cognitive Sciences, 7*(12), 527-533.

Dittrich, W. H. (1993). Action Categories and the Perception of Biological Motion. *Perception, 22*(1), 15-22.

Durgin, F. H., & Li, Z. (2010). Controlled interaction: Strategies for using virtual reality to study perception. *Behavior Research Methods, 42*(2), 414-420.

Farrow, D., & Abernethy, B. (2003). Do expertise and the degree of perception - action coupling affect natural anticipatory performance? *Perception, 32*(9), 1127-1139.

Fink, P. W., Foo, P. S., & Warren, W. H. (2009). Catching fly balls in virtual reality: A critical test of the outfielder problem. *Journal of Vision, 9*(13)

Flavell, J. H. (2004). Development of knowledge about vision. *Thinking and Seeing*, 13-36.

Fodor, J. A. (1987). *Psychosemantics : the problem of meaning in the philosophy of mind.* Cambridge, Mass. ; London: published in cooperation with the British Psychological Society [by] MIT Press.

Frith, C. D., & Frith, U. (2006). The neural basis of mentalizing. *Neuron, 50*(4), 531-534.

Frith, U., & Frith, C. (2010). The social brain: allowing humans to boldly go where no other species has been. *Philosophical Transactions of the Royal Society B-Biological Sciences, 365*(1537), 165-175.

Gallese, V., Fadiga, L., Fogassi, L., & Rizzolatti, G. (1996). Action recognition in the premotor cortex. *Brain, 119*, 593-609.

Gallese, V., & Goldman, A. (1998). Mirror neurons and the simulation theory of mind-reading. *Trends in Cognitive Sciences, 2*(12), 493-501.

Georgiou, I., Becchio, C., Glover, S., & Castiello, U. (2007). Different action patterns for cooperative and competitive behaviour. *Cognition, 102*(3), 415-433.

Grezes, J., Frith, C., & Passingham, R. E. (2004a). Brain mechanisms for inferring deceit in the actions of others. *Journal of Neuroscience, 24*(24), 5500-5505.

Grezes, J., Frith, C. D., & Passingham, R. E. (2004b). Inferring false beliefs from the actions of oneself and others: an fMRI study. *Neuroimage, 21*(2), 744-750.

Grierson, L. E. M., Gonzalez, C., & Elliott, D. (2009). Kinematic Analysis of Early Online Control of Goal-Directed Reaches: A Novel Movement Perturbation Study. *Motor Control, 13*(3), 280-296.

Houlston, D. R., & Lowes, R. (1993). Anticipatory Cue-Utilization Processes Amongst Expert and Nonexpert Wicketkeepers in Cricket. *International Journal of Sport Psychology, 24*(1), 59-73.

Huys, R., Canal-Bruland, R., Hagemann, N., Beek, P. J., Smeeton, N. J., & Williams, A. M. (2009). Global Information Pickup Underpins Anticipation of Tennis Shot Direction. *Journal of Motor Behavior, 41*(2), 158-170.

Iacoboni, M., Woods, R. P., Brass, M., Bekkering, H., Mazziotta, J. C., & Rizzolatti, G. (1999). Cortical mechanisms of human imitation. *Science, 286*(5449), 2526-2528.

Isaacs, L. D., & Finch, A. E. (1983). Anticipatory Timing of Beginning and Intermediate Tennis Players. *Perceptual and Motor Skills, 57*(2), 451-454.

Johansson. (1973). Visual-Perception of Biological Motion and a Model for Its Analysis. *Attention, Perception, & Psychophysics, 14*(2), 201-211.

Johansson, R. S., Westling, G. R., Backstrom, A., & Flanagan, J. R. (2001). Eye-hand coordination in object manipulation. *Journal of Neuroscience, 21*(17), 6917-6932.

Keller, P. E., Knoblich, G., & Repp, B. H. (2007). Pianists duet better when they play with themselves: On the possible role of action simulation in synchronization. *Consciousness and Cognition, 16*(1), 102-111.

Kelso, J. A. (1984). Phase transitions and critical behavior in human bimanual coordination. *Am J Physiol, 246*, R1000-1004.

Knoblich, G., & Sebanz, N. (2008). Evolving intentions for social interaction: from entrainment to joint action. *Philosophical Transactions of the Royal Society B-Biological Sciences, 363*(1499), 2021-2031.

Kording, K. P., Fukunaga, I., Hovard, I. S., Ingram, J. N., & Wolpert, D. M. (2004). A neuroeconomics approach to inferring utility functions in sensorimotor control. *Plos Biology, 2*(10), 1652-1656.

Lagarde, J., & Kelso, J. A. S. (2006). Binding of movement, sound and touch: multimodal coordination dynamics. *Experimental Brain Research, 173*(4), 673-688.

Lee, D. N., Young, D. S., Reddish, P. E., Lough, S., & Clayton, T. M. H. (1983). Visual Timing in Hitting an Accelerating Ball. *Quarterly Journal of Experimental Psychology Section a-Human Experimental Psychology, 35*(May), 333-346.

Loomis, J. M., Blascovich, J. J., & Beall, A. C. (1999). Immersive virtual environment technology as a basic research tool in psychology. *Behavior Research Methods Instruments & Computers, 31*(4), 557-564.

Mann, D. L., Abernethy, B., & Farrow, D. (2010). Action specificity increases anticipatory performance and the expert advantage in natural interceptive tasks. *Acta Psychologica, 135*(1), 17-23.

Manstead, A. S. R., & Hewstone, M. (1996). *The Blackwell encyclopedia of social psychology*: Wiley-Blackwell.

McBeath, M. K., Shaffer, D. M., & Kaiser, M. K. (1995). How Baseball Outfielders Determine Where to Run to Catch Fly Balls. *Science, 268*(5210), 569-573.

McLeod, P., & Dienes, Z. (1993a). Running to catch the ball. *Nature, 362*(6415), 23.

McLeod, P., & Dienes, Z. (1996). Do fielders know where to go to catch the ball or only how to get there? *Journal of Experimental Psychology-Human Perception and Performance, 22*(3), 531-543.

McLeod, P., Reed, N., & Dienes, Z. (2006). The generalized optic acceleration cancellation theory of catching. *Journal of Experimental Psychology-Human Perception and Performance, 32*(1), 139-148.

McLeod, P., Reed, N., Gilson, S., & Glennerster, A. (2008). How soccer players head the ball: A test of optic acceleration cancellation theory with virtual reality. *Vision Research, 48*(13), 1479-1487.

McMorris, T., & Colenso, S. (1996). Anticipation of professional soccer goalkeepers when facing right- and left-footed penalty kicks. *Perceptual and Motor Skills, 82*(3), 931-934.

McNeill, W. H. (1995). *Keeping together in time : dance and drill in human history.* Cambridge, Ma: Harvard University Press.

Milner, A. D., & Goodale, M. A. (1995). *The visual brain in action.* Oxford: Oxford University Press.

Milner, A. D., & Goodale, M. A. (2008). Two visual systems re-viewed. *Neuropsychologia, 46*(3), 774-785.

Neda, Z., Ravasz, E., Brechet, Y., Vicsek, T., & Barabasi, A. L. (2000). The sound of many hands clapping - Tumultuous applause can transform itself into waves of synchronized clapping. *Nature, 403*(6772), 849-850.

Ochsner, K. N., & Lieberman, M. D. (2001). The emergence of social cognitive neuroscience. *American Psychologist, 56*(9), 717-734.

Oullier, O., De Guzman, G. C., Jantzen, K. J., Lagarde, J., & Kelso, J. A. S. (2008). Social coordination dynamics: Measuring human bonding. *Social Neuroscience, 3*(2), 178-192.

Penrose, J. M. T., & Roach, N. K. (1995). Decision making and advanced cue utilisation by cricket batsmen. *Journal of Human Movement Studies, 29*(5), 199-218.

Pikovsky, A., Rosenblum, M., & Kurths, J. (2001). *Synchronization : a universal concept in nonlinear sciences.* Cambridge: Cambridge University Press.

Poulton, E. C. (1957). On prediction in skilled movements. *Psychol Bull, 54*(6), 467-478.

Premack, D., & Woodruff, G. (1978). Does the Chimpanzee Have a Theory of Mind. *Behavioral and Brain Sciences, 1*(4), 515-526.

Prinz, W. (1984). Modes of linkage between perception and action. *Cognition and motor processes*, 185-193.

Prinz, W., & Hommel, B. (2002). Common mechanisms in perception and action - Introductory remarks. *Common Mechanisms in Perception and Action, 19*, 3-5.

Proteau, L., Boivin, K., Linossier, S., & Abahnini, K. (2000). Exploring the limits of peripheral vision for the control of movement. *Journal of Motor Behavior, 32*(3), 277-286.

Reichenbach, A., Thielscher, A., Peer, A., Bülthoff, H. H., & Bresciani, J. P. (2009). Seeing the hand while reaching speeds up on-line responses to a sudden change in target position. *Journal of Physiology-London, 587*(19), 4605-4616.

Renshaw, I., & Fairweather, M. M. (2000). Cricket bowling deliveries and the discrimination ability of professional and amateur batters. *J Sports Sci, 18*(12), 951-957.

Richardson, M. J., Marsh, K. L., Isenhower, R. W., Goodman, J. R. L., & Schmidt, R. C. (2007). Rocking together: Dynamics of intentional and unintentional interpersonal coordination. *Human Movement Science, 26*(6), 867-891.

Richardson, M. J., Marsh, K. L., & Schmidt, R. C. (2005). Effects of visual and verbal interaction on unintentional interpersonal coordination. *Journal of Experimental Psychology-Human Perception and Performance, 31*(1), 62-79.

Rizzolatti, G., & Sinigaglia, C. (2010). The functional role of the parieto-frontal mirror circuit: interpretations and misinterpretations. *Nature Reviews Neuroscience, 11*(4), 264-274.

Runeson, S., & Frykholm, G. (1983). Kinematic Specification of Dynamics as an Informational Basis for Person-and-Action Perception - Expectation, Gender Recognition, and Deceptive Intention. *Journal of Experimental Psychology-General, 112*(4), 585-615.

Ruys, K. I., & Aarts, H. (2010). When competition merges people's behavior: Interdependency activates shared action representations. *Journal of Experimental Social Psychology, 46*(6), 1130-1133.

Sailer, U., Flanagan, J. R., & Johansson, R. S. (2005). Eye-hand coordination during learning of a novel visuomotor task. *Journal of Neuroscience, 25*(39), 8833-8842.

Salmela, J. H., & Fiorito, P. (1979). Visual cues in ice hockey goaltending. *Can J Appl Sport Sci, 4*(1), 56-59.

Sarlegna, F. R., & Blouin, J. (2010). Visual guidance of arm reaching: Online adjustments of movement direction are impaired by amplitude control. *Journal of Vision, 10*(5)

Savelsbergh, G. J. P., Williams, A. M., Van der Kamp, J., & Ward, P. (2002). Visual search, anticipation and expertise in soccer goalkeepers. *Journal of Sports Sciences, 20*(3), 279-287.

Schilbach, L., Timmermans, B., Reddy, V., Costall, A., Bente, G., Schlicht, T., & Vogeley, K. (2012). Toward a second-person neuroscience. *Behav. Brain Res.*

Schmidt, R. C., Carello, C., & Turvey, M. T. (1990). Phase-Transitions and Critical Fluctuations in the Visual Coordination of Rhythmic Movements between People. *Journal of Experimental Psychology-Human Perception and Performance, 16*(2), 227-247.

Sebanz, N., Bekkering, H., & Knoblich, G. (2006). Joint action: bodies and minds moving together. *Trends in Cognitive Sciences, 10*(2), 70-76.

Sebanz, N., & Knoblich, G. (2009). Prediction in Joint Action: What, When, and Where. *Topics in Cognitive Science, 1*(2), 353-367.

Sebanz, N., & Shiffrar, M. (2009). Detecting deception in a bluffing body: The role of expertise. *Psychonomic Bulletin & Review, 16*(1), 170-175.

Sellars, W. (1956). Empiricism and the Philosophy of Mind. *Minnesota Studies in the Philosophy of Science, 1*, 253-329.

Shockley, K., Santana, M. V., & Fowler, C. A. (2003). Mutual interpersonal postural constraints are involved in cooperative conversation. *Journal of Experimental Psychology-Human Perception and Performance, 29*(2), 326-332.

Spijkers, W., & Spellerberg, S. (1995). On-line visual control of aiming movements? *Acta Psychol (Amst), 90*(1-3), 333-348.

Streuber, S., Knoblich, G., Sebanz, N., Bülthoff, H. H., & de la Rosa, S. (2011). The effect of social context on the use of visual information. *Experimental Brain Research, 214*(2), 273-284.

Triplett, N. (1898). The dynamogenic factors in pacemaking and competition. *The American journal of psychology, 9*(4), 507--533.

Turvey, M. T. (1990). Coordination. *American Psychologist, 45*(8), 938-953.

Urgesi, C., Savonitto, M. M., Fabbro, F., & Aglioti, S. M. (2011). Long- and short-term plastic modeling of action prediction abilities in volleyball. *Psychol Res.*

Van Avermaet, E. (1996). Cooperation and competition. *The Blackwell Handbook of Social Psychology*, 136--141.

Van der Kamp, J., Rivas, F., Van Doorn, H., & Savelsbergh, G. (2008). Ventral and dorsal system contributions to visual anticipation in fast ball sports. *International Journal of Sport Psychology, 39*(2), 100-130.

Vanrie, J., & Verfaillie, K. (2004). Perception of biological motion: A stimulus set of human point-light actions. *Behavior Research Methods Instruments & Computers, 36*(4), 625-629.

Vesper, C., Butterfill, S., Knoblich, G., & Sebanz, N. (2010). A minimal architecture for joint action. *Neural Networks, 23*(8-9), 998-1003.

Vignais, N., Bideau, B., Craig, C., Brault, S., Multon, F., Delamarche, P., & Kulpa, R. (2009). Does the level of graphical detail of a virtual handball thrower influence a goal-keeper's motor response? *Journal of Sports Science and Medicine, 8*(4), 501-508.

Vignais, N., Kulpa, R., Craig, C., Brault, S., Multon, F., & Bideau, B. (2010). Influence of the graphical levels of detail of a virtual thrower on the perception of the movement. *Presence: Teleoperators and Virtual Environments, 19*, 243-252.

Wann, J. P., & Rushton, S. K. (1995). The use of virtual environments in perception action research: Grasping the impossible and controlling the improbable. *Motor Control and Sensory Motor Integration, 111*, 341-360.

Ward, P., Williams, A. M., & Bennett, S. J. (2002). Visual search and biological motion perception in tennis. *Res Q Exerc Sport, 73*(1), 107-112.

Warren, W. H. (2006). The dynamics of perception and action. *Psychological Review, 113*(2), 358-389.

Warren, W. H., Kay, B. A., Zosh, W. D., Duchon, A. P., & Sahuc, S. (2001). Optic flow is used to control human walking. *Nature Neuroscience, 4*(2), 213-216.

Weissensteiner, J., Abernethy, B., Farrow, D., & Muller, S. (2008). The Development of Anticipation: A Cross-Sectional Examination of the Practice Experiences Contributing to Skill in Cricket Batting. *Journal of Sport & Exercise Psychology, 30*(6), 663-684.

Williams, A. M., Davids, K., & Williams, J. G. P. (1999). *Visual perception and action in sport*. New York ; London: E&FN Spon.

Williams, A. M., Ward, P., Knowles, J. M., & Smeeton, N. J. (2002). Anticipation skill in a real-world task: measurement, training, and transfer in tennis. *J Exp Psychol Appl, 8*(4), 259-270.

Winfree, A. T. (2002). Oscillating systems. On emerging coherence. *Science, 298*(5602), 2336-2337.

Wolpert, D. M., Ghahramani, Z., & Flanagan, J. R. (2001). Perspectives and problems in motor learning. *Trends in Cognitive Sciences, 5*(11), 487-494.

Christian M. Stracke

Advances in Irrigation and Hydroponics
Competence & Skills Development in Agriculture & Aquaculture

AGRICOM is the innovative research project of Lifelong Learning Programme funded by the European Agency for Education and Culture (EACEA) and managed by the National Agency in Germany (NA BIBB).

UNIVERSITÄT
DUISBURG
ESSEN

Open-Minded

Christian M. Stracke (Ed.)

Advances in Irrigation and Hydroponics
Competence & Skills Development in
Agriculture & Aquaculture

Bibliographic information published by the Deutsche Nationalbibliothek

The Deutsche Nationalbibliothek lists this publication in the Deutsche Nationalbibliografie; detailed bibliographic data are available in the Internet at http://dnb.d-nb.de

Published by Logos Verlag Berlin GmbH
Comeniushof, Gubener Str. 47,
10243 Berlin
Tel.: +49 (0)30 42 85 10 90
Fax: +49 (0)30 42 85 10 92
INTERNET: http://www.logos-verlag.de

The digital copy of this version is online available: http://www.agriculture-competences.eu

Contact:
Christian M. Stracke
University of Duisburg-Essen
Universitaetsstr. 9
45141 Essen, Germany
christian.stracke@uni-due.de

Information about AGRICOM project online: http://www.agriculture-competences.eu

Information about the University of Duisburg-Essen online: http://www.wip.uni-due.de

ISBN: 978-3-8325-3540-7

Content

Advances in Irrigation and Hydroponics Competence & Skills Development in Agriculture & Aquaculture
An Introduction

Due to the shortage of the precious resource water and the consequences of the climate change, numerous European policies have been developed and adopted for the protection and sustainable utilisation of water creating a huge demand in particular in the vocational training. Economic factors like privatisation and increasing cost pressure in water management are sharpening these educational needs leading to the demand for specific Vocational Education and Training (VET) opportunities and products as short and tailor-made as possible.

The AGRICOM project supports close links to working life in order to make VET more responsive to the labour market's needs in the agricultural sector. AGRICOM facilitates and improves the identification and anticipation of skills and competences' needs and their integration in VET provision and implies also promoting integration of learning with working. In particular AGRICOM supports the implementation of the "New Skills for New Jobs" strategy by taking into account the challenges such as environmental and demographic changes and the related growing job needs also in the agricultural sector.

The presented articles are the result of the Open Call for Papers issued by the AGRICOM Conference, which took place in Viterbo on 19th and 20th of September 2013 under the title: "Advances in Irrigation and Hydroponics: Competence & Skills Development in Agriculture & Aquaculture".

The AGRICOM Consortium has organised the AGRICOM Conference in order to raise the awareness of the stakeholders in the agricultural sector concerning competence modelling, with a special focus on hydroponics and irrigation. Thus, the AGRICOM Conference aimed to provide an opportunity for meetings and discussions between scientists and experts from all over Europe who are either subject experts or who deal with competence modelling as well as educational and training issues in the field of agriculture.

The scientific articles published in this book are the selected papers of applicants from over six countries received upon the Open Call for Papers issued by the AGRICOM Conference 2013: They were reviewed by the scientific Programme Committee of AGRICOM 2013 in double-blind peer reviews and selected according the review results. In addition all authors of the selected articles could present and discuss their papers at the AGRICOM conference in a speech.

Daniela Borgognone, Eva Švecová, Antonio Fiorillo, Elvira Rea, Mariateresa Cardarelli, Giuseppe Colla (all from Italy) present the effect of irrigation regime on growth, flowering and water use of Bougainvillea spectabilis. The effect of three irrigation regimes on morpho-physiological parameters and product quality of Bougainvillea spectabilis was investigated with an emphasis on the tolerance to water stress in post-production. The results showed that the use of an irrigation regime based on 60% of daily water use can be recommended for potted production of Bougainvillea because it allows to reduce water consumption and to increase water use efficiency without any detrimental effect on plant growth.

Andrea Dührkoop (Germany), Prof. Dr. Tarik Hartani (Algeria), Edward Muchiri (Algeria), Abdelaali Bencheikh (Algeria), Tarek A. Ouamane (Algeria), Madjed A. Djoudi ((Algeria), Prof. Dr. Oliver Hensel (Germany) present an interdisciplinary research project, aiming at the modernisation of pitcher-irrigation by applying modern materials and production technology based on semi-permeable membranes accompanied by trainings for the involved local scientific staff.

Manuela Capodilupo and Accursio Venezia (both from Italy) demonstrate current research in trough bench subirrigation system for tomato and other vegetables initiated by the OFRALSER project.

A. Parente (Italy), Y. L. Tsirogiannis (Greece), M. Nicolaos (Greece), B. Pantelis (Greece) and F. Montesano (Italy) present a project aiming at the establishment of a network of knowledge and expertise which will lead to the development of practical irrigation management tools for demand driven capitalization of scientific knowledge and good practices: "The IRMA project: Efficient Irrigation Management Tools for Agricultural Cultivations and Urban Landscapes".

Andrea Dührkoop and Prof. Dr. Oliver Hensel (both from Germany) present a joint research program to establish links between projects dealing with subsurface irrigation (SI) methods in Ivory Coast, Algeria, Kenya, Turkey and Namibia. Each country will conduct an analysis of SI methods in field

trials under local conditions in terms of water productivity, use of low quality water, salinity effects in the soil and socio economic aspects. to summarize the results into guidelines for stakeholders, farmers, water managers, water authorities and ministries.

Cornelia Helmstedt, Xenia Rodriguez and Christian M. Stracke (all from Germany) present the AGRICOM Project which main goal is to establish the first Competence Model for the Agricultural Sector (ACM) in order to strengthen the transparency and comparability of VET opportunities at a European level.

Henrike Perner (from Germany) describes the method used to transfer the water competence model (WCM) into the agricultural and horticultural sector in the framework of the AGRICOM project.

Andreas Drakos, Charalampos Thanopoulos, Yannis Psochios (all from Greece) introduce to one of the main outputs of the AGRICOM initiative: the web portal and repository for storing and hosting Vocational Education and Training (VET) elements based on the AGRICOM Competence Model (ACM) and the implemented quality management system. The system forces uploaded content through a specific workflow that requires user interaction and validation before publishing.

To summarise, this book contributes to the current developments and debate on competence modelling in the agricultural and horticultural sectors by presenting latest techniques, offering different views and solutions on competence modelling and by providing suggestions for future improvements European vocational education and training in the agricultural and horticultural sectors.

Cornelia Helmstedt, Xenia Rodriguez and Christian M. Stracke

Effect of irrigation regime on growth, flowering and water use of Bougainvillea spectabilis

Daniela Borgognone[1], Eva Švecová[1], Antonio Fiorillo[1], Elvira Rea[2], Mariateresa Cardarelli[2], Giuseppe Colla[1]

[1] Dipartimento di scienze e tecnologie per l'Agricoltura, le Foreste, la Natura e l'Energia, Università della Tuscia, via San Camillo De Lellis snc, 01100 Viterbo, Italy.
[2] Consiglio per la Ricerca e la sperimentazione in Agricoltura – Centro di Ricerca per lo studio delle relazioni tra Pianta e Suolo, via della Navicella 2-4, 00184 Roma, Italy.

Abstract. *Bougainvillea* Comm. ex Juss. (*Nyctaginaceae*) is an important ornamental plant widely grown in mild and warm regions. The selection of varieties and good management of the cultivation reflect on the performance of plants in post-production. During the transport, Bougainvillea plants are often subjected to a water stress leading to physiological disorders and severe damages of plant material. To avoid the damages by water stress, the increase of drought tolerance is essential. In the present study, the effect of three irrigation regimes on morpho-physiological parameters and product quality of *Bougainvillea spectabilis* was investigated with an emphasis on the tolerance to water stress in post-production. Leaf water potential and flowering were significantly affected by irrigation regime. Increasing irrigation amount increased water consumption and decreased water use efficiency. The results showed that the use of an irrigation regime based on 60% of daily water use can be recommended for potted production of Bougainvillea because it allows to reduce water consumption and to increase water use efficiency without any detrimental effect on plant growth.

Keywords: Bougainvillea, irrigation regime control, water stress, drought tolerance.

1 Introduction

Bougainvillea Comm. ex Juss. (*Nyctaginaceae*) is an important genus from the category of ornamental plants, widely grown in mild and warm regions. It includes numerous species of arborescent shrubs and climbing plants with origin in the tropical regions of South America (Huxley et al., 1999). The species of greatest interest are the *B. spectabilis*, *B. glabra* and *B. x buttiana* (Dole and Wilkins, 1999). Each species contains several subspecies characterized by different growth habit, bract colour and the foliage (Huxley et al., 1999). Thanks to the remarkable variety of forms and strong adaptability to the Mediterranean climatic conditions, bougainvillea has spread widely in Italy (Leonardi and Romano, 2003).

Due to the great demand, this ornamental plant requires the appropriate management of the large scale plant production, and subsequently the adequate logistics. In fact, during the transport Bougainvillea plants are often subjected to a water stress leading to various physiological disorders. Although in some cases the controlled water stress is required to control plant height and width, e.g. in *B. spectabilis* (Ma and Gu, 2012), uncontrolled water stress leads to severe damages of plant material. Currently, little is known about the physiological disorders occurring to the potted plants under water stress conditions.

Drought tolerance might be increased by irrigation regime control (Fernández et al., 2006). In the study of Li and Yang (2009), the water use efficiency of *Bougainvillea glabra* was highest after four days of water stress, indicating that moderate water stress improves the water use efficiency. These findings suggest that pre-adaptation of the plants to a drought stress may increase resistance to the stress during the transport.

The aim of the present study was to investigate the effect of three irrigation regimes on growth, flowering, and water use efficiency of *Bougainvillea spectabilis*.

2 Materials and Methods

The trial was carried out at the Experimental Farm of Tuscia University, central Italy (lat. 42°25′N, long. 12°08′E, alt. 310 m above sea level) under greenhouse conditions. The greenhouse was maintained at day/night temperatures between 14 and 28°C, and day/night relative humidity of 50/85%. Rooted cuttings of *Bougainvillea spectabilis* 'Fucsia colour' from Torsanlorenzo Nursery, Torsanlorenzo (RM), Italy were transplanted on 1 August 2010 in plastic pots (diameter 20 cm) filled with a mixture of peat/pumice (2:1 v/v) substrate.

Three irrigation regimes were tested in a randomized complete block design with four replicates. Irrigation regimes were obtained by varying the level of water recovery (100%, 60% and 40% of the daily water use obtained respectively with drippers 8, 6 and 2 L/h). Daily water use was equal to the water required to bring the substrate to container capacity plus 15% of runoff. Plants were fertigated prior planting with a slow release fertilizer (4 g/L of Baycote, Bayer) and during the cropping cycle with a complete nutrient solution.

On 24 September 2010 (55 days from transplanting, DAT), plant height, number of stems per plant and SPAD index of leaves were measured. Moreover, on 6 January 2011 (159 DAT), SPAD index of leaves was measured. At the end of the cropping cycle (257 DAT), leaf water potential, SPAD index of leaves, percentage of flowering, fresh weight of leaves, stems, and inflorescence were measured. Leaves, stems, and inflorescences were dried in a forced-air oven at 80 °C for 72 h for dry biomass determination. Water consumption was monitored during the cropping cycle using a gravimetric method. Water use efficiency was calculated as the ratio between above ground dry biomass and water consumption.

3 Results

At 55 DAT, no significant differences were recorded on plant height and number of stems among plants irrigated with different water amounts (avg. 102.6 cm and 2.8 stems/plant, respectively). No significant differences were

recorded on SPAD of leaves during the cropping cycle (avg. 57.4, 70.9, 74.9 at 55, 159, and 257 DAT, respectively). At the end of the cropping cycle, leaf water potential was significantly affected by irrigation regime with higher value in leaves of plants irrigated with 40% of DWU (-1.69 MPa) than in those irrigated with 100% DWU (-1.22 MPa) while leaves of plants irrigated with 60% DWU gave an intermediate value (-1.48 MPa) Total above ground dry biomass was significantly lower in 40% DWU treatment than in 100% and 60% DWU (88 vs 175 g/plant).

Water regime significantly affected the flowering (P<0.01) with higher value in plants irrigated with 100 and 60% of daily water use (avg 22.6%) compared to plants irrigated with 40% of DWU (16.1%). Increasing irrigation amount increased water consumption (Fig. 1) and decreased water use efficiency (1.1, 1.8, and 2.1 g/L for 100, 60 and 40% of DWU, respectively). Similar results were achieved in wheat, where an increased irrigation regime caused decrease in water use efficiency and lead to increase in days to maturity and water use (Onyibe, 2005). Similarly, water use efficiency significantly increased under water stress in *Catharanthus roseus* (L.) G. Don. (Jaleel et al., 2005). These plant responses suggest that controlled water stress might contribute to stress tolerance, that is essential especially during the transport, and to reduce the water use during the growth cycle at the same time.

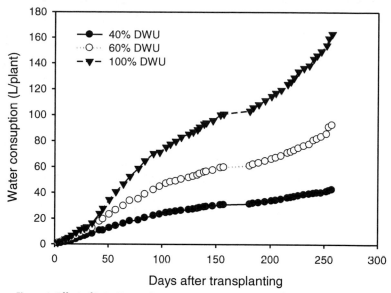

Figure 1: Effect of irrigation regimes on water consumption of Bougainvillea plants.

4 Conclusion

The results demonstrated that the use of an irrigation regime based on 60% of daily water use can be recommended for potted production of Bougainvillea because it allows to reduce water consumption and to increase water use efficiency without any detrimental effect on plant growth.

5 Acknowledgment

This work was financed by Ministry of Agricultural, Food and Forestry Policies (MiPAAF) Project "Tecnologie di filiera per il controllo della tolleranza a stress idrico in Bougainvillea"(D.M. 11053/7643/09 of 7 May 2009).

6 References

Dole, J. M., Wilkins, H. F., 1999. Floricolture principles and species. 1st ed. Prentice Hall, Upper Saddle River, N. J.

Fernàndez, J. A., Balenzategui, L., Banon, S., Franco, J. A., 2006. Induction of drought tolerance by paclobutrazol and irrigation deficit in *Phillyrea angustifolia* durino nursery period. *Scientia Horticulturae* 107: 277-283.

Huxley A., Griffths M., Levy M., 1999. The New Royal Horticultural Society Dictionary of gardening. The Macmillan Press Limited, Londra.

Jaleel, C. A., Gopi R., Sankar, B., Gomathinayagam, M., Panneerselvam, R., 2008. Differential responses in water use efficiency in two varieties of *Catharanthus roseus* under drought stress. *C R Biol.* 2008 Jan;331(1):42-7.

Leonardi C., Romano D., 2003. Caratteristiche di tipi di *Bougainvillea* diffusi nella Sicilia orientale. *Italus Hortus*, 10: 234-237.

Li, Y.-h.,Yang, Y., 2009. The Effect of Water Stresses to Photosynthesis of *Bougainvillea glabra*. *Northern Horticulture*, 1, p. 181.

Ma, S., Gu, M., 2012.Effects of water stress and selected plant growth retardants on growth and flowering of 'raspberry ice' bougainvillea *(Bougainvillea spectabilis)*. *Acta Horticulturae*, Volume 937, 30 September 2012, 237-242.

Onyibe, J. E., 2005. Effect of irrigation regime on growth and development of two wheat cultivars *(Triticum aestivum* L.) in the Nigerian savanna. *Journal of Agriculture and Rural Development in the Tropics and Subtropics*, 106 (2), 177-192.

International research activities on innovative low pressure irrigation technique with polymer membrane

Andrea Dührkoop[1], Prof. Dr. Tarik Hartani[2], Edward Muchiri[3], Abdelaali Bencheikh[4], Tarek A. Ouamane[4], Madjed A. Djoudi[4], Prof. Dr. Oliver Hensel[1]

[1] Kassel University, Department for Agricultural Engineering, Nordbahnhofstr. 1a, 37213 Witzenhausen, Germany.
[2] National Advanced School of Agronomy (ENSA), Institute for Water in Agriculture, Algeria.
[3] Egerton University, Department of Civil & Environmental Engineering, Kenya.
[4] Technical and Scientific Research Center on Arid Regions (CRSTRA), Biskra, Algeria.

Abstract: The paper presents an interdisciplinary and effect-oriented research project, aiming at the development of a both ecological and economical irrigation technology. The research project is financed by the German Federal Ministry for Education and Research (BMBF) within the framework of the support programme "Sustainable Solutions for Sub-Saharan Africa". The involved project partners comprise Universities, research institutes and industry companies from Algeria, Kenya and Germany.

The technological approach is focused on a modernization of pitcher-irrigation. This irrigation method - well known for its superior efficiency in arid areas since ancient times - has the disadvantage of not to be economic when used in modern agriculture. The project adopts the pitcher-irrigation principle to modern materials and production technology based on semi-permeable membranes. Field tests in Algeria and Kenya are presented, showing promising results with respect to economic and ecologic aspects.

Keywords: irrigation, subsurface irrigation, water efficiency, auto-regulative, water saving, intelligent irrigation.

1 Introduction

In Algeria, Kenya and many other states of arid and semi-arid zones, water consumption exceeds the renewable water resources, leading to widespread groundwater depletion and water scarcity. Considering climate change and poverty as well as water and food scarcity it is vital to enhance crop yields while at the same time reducing water demand. Current irrigation methods use only a minor portion of the applied water, and up to 60 % of water applied is lost due to percolation, evaporation and poor water management (Hübener, 2006).

Within this context the BMBF has initiated joint research activities among Universities, industry companies and research institutes from Germany and Subsahara Africa financed for a period of 24 months. The presented project comprises two key aspects: the first is the testing and development of an innovative auto-regulative subsurface irrigation technique with membrane material and the second is the qualification of the involved project partners and their junior researchers in order to intensify research cooperation between Germany and countries from Subsahara Africa.

2 Project structure

The project was initiated by the Institute of Agricultural Engineering of the Kassel University in Germany, having a long experience in irrigation, especially in subsurface and pitcher irrigation. Due to the institute's numerous research and education activities in Africa partners for the BMBF project were easy to find. Preliminary studies carried out at the institute could be transferred to field conditions in the partner countries Algeria and Kenya.

On the African partner side Algerian and Kenyan four research institutes are involved. From Algeria there are:

- the National Advabced School of Agronomy (ENSA), Agricultural water management laboratory – Institute for Water in Agriculture

- the Research Center for Science and Technologies in Arid regions (C.R.S.T.R.A.) and
- the Technical Institute of Vegetable and Industrial Culture – Ministry of Agriculture (ITCMI).

From the Kenyan side there is the University of Egerton with its department of Civil and Environmental Engineering.

The pool of involved project partners comprises on the German side:

- the Department of Agricultural Engineering of the Kassel University,
- the Institute for Tropical and Subtropical Agriculture (DITSL),
- Deutsche Gesellschaft für Internationale Zusammenarbeit (GIZ) GmbH with its Algerian Integrated Water Resources Program,
- the Chair of Chemical Process Engineering of the RWTH Aachen.

Partners from industry comprise MICRODYN-NADIR (Wiesbaden, Membrane Technology) and PFAFF Industrial (Kaiserslautern, welding technology for synthetic tissues), furnishing know-how and materials necessary for prototype production.

3 Scientific objectives and results

The project bundles research and development activities of various stakeholders in order to develop a new environment friendly and water saving technology.

Within the cooperation period the following research activities were accomplished:

A. Developing and testing of the water saving subsurface irrigation technology.

The used pipe for irrigation purposes in this project is normally used for medical or filtration applications. Its functionality for an auto-regulative

subsurface irrigation application can be assumed due to its hydraulic properties. Field tests confirmed results of preliminary laboratory testing.

As with the traditional pitcher irrigation, the irrigation pipe is characterized by its ability of auto regulation of water flow, which arises from the close interaction between the plant, the dry soil and the pipe material. This feature makes the system very efficient, as water flows out of the irrigation pipe only when the soil is dry and the crop suffers water stress. If the soil humidity rises, the water flow will decrease or even stop automatically.

Experiments are actually conducted on Algerian test sites from ENSA and C.R.S.T.R.A. and on Kenyan test fields of the Egerton University to assess the auto-regulative system performance at three study sites (Algiers and Biskra in Algeria and Nakuru in Kenya). The irrigation system was installed at 30 cm depth below tomato plants grown in greenhouses. Each experimental trial was replicated and compared to a common drip irrigation system.

The following photos give an impression of the irrigation trial in a greenhouse at the experimental station El Outaya of the Research Center for Science and Technologies in Arid regions (C.R.S.T.R.A.) in Algeria.

Irrigation pipe

Installation of the irrigation pipe in a greenhouse trial

Andrea Dührkoop, Prof. Dr. Tarik Hartani, Edward Muchiri, Abdelaali Bencheikh, Tarek A. Ouamane, Madjed A. Djoudi, Prof. Dr. Oliver Hensel

Installation depth 30 cm

Connection with the submain line (PE pipe ∅ 17 mm)

Planted crop: tomatoes

Dry soil surface

Water consumption and tomato production were calculated with respect to various parameters: climate, water quality and soil characteristics. In addition to high quality well water, saline and treated wastewaters were tested. Soils characteristics varied between clay, loamy-clay and sandy clay.

First results of the investigations confirm the assumption that water consumption with the auto-regulative system is lower than with drip irrigation (see figure 1, results of a greenhouse trial by the Algerian partner ENSA, the water consumption is for 156 tomato plants on a surface of 104 m²). In addition, the water use efficiency defined as units of produced fruit per amount of delivered water is higher with the auto-regulative system.

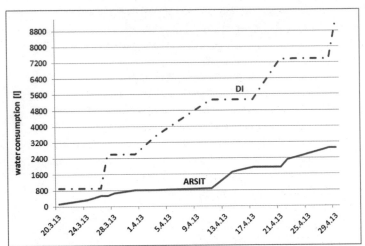

Figure 1: Cumulated water consumption, greenhouse trial of the project partner ENSA,
ARSIT = auto regulative subsurface irrigation technology, DI = drip irrigation

The developed auto-regulative irrigation system does not require a monitoring, inhibits weeds because of the dry soil surface and allows the use of liquid fertilizer.

B. Investigations of the retention capacity of the polymer membrane against dissolved mineral salts and of salinization effects.

Soil degradation caused by salt is a problem common even to advanced localized irrigation methods, such as drip irrigation. Investigation of salt distribution around the sub-soil irrigation pipe is of major importance in the research project.

The irrigation method developed in this project is less affected by salinization, as evaporation has little influence on the water applied. See figure 2 where a buried membrane pipe is shown. The soil around the pipe is wet but the soil surface is dry. So the influence of evaporation on the subsurface system is negligible and no losses by evaporation can arise.

Figure 2: Dry soil surface over the irrigation pipe

Further detailed results will be published in future publications.

C. Analysis of the suitability for low quality irrigation water

In order to preserve natural water resources, the use of treated waste water or water with low quality is desirable. Regarding this aspect it was investigated if the polymer-membrane pipe wall of the irrigation pipe is capable to deal with low-quality water.

First trials with treated waste water were conducted at the campus of the Egerton University in Kenya in 2013. The results will be available soon.

A trial with harvested rain water has shown that the pipe was totally clogged by solid particles (especially silty soil fractions). The diameter of the tested membrane pipe is very small so it is easy to clog the pipe by bigger particles. For a good functioning system operation a pipe flushing modus must be consider.

D. Identify barriers for the implementation of the new irrigation technology.

This part of the program, which is focused on socio-economic aspects, has also the ambition to identify barriers to implementation of the new water-saving technology. It will propose recommendations to improve the dialogue between the different involved parties, and provide information to the involved stakeholders (farmers, water managers, agricultural advisors, etc.). These studies are mainly conducted by colleagues from CRSTRA (Algeria), results will be given in future publications.

4 Qualification measures

Project workshops and the supervision of Master thesis or doctoral thesis form part of the project. One workshop was conducted at the Algerian partner C.R.S.T.R.A. in Biskra, where the involved partners, local administrations and farmers participated. The main discussion points were the utilisation of the auto-regulative subsurface irrigation technique by the farmers and the functionality of the innovative system in comparison with common used irrigation systems.

With presentations and round tables an engaged communication was opened between the different participants.

The photos show experiences of the workshop in Biskra in 2013 at C.R.S.T.R.A.

Presentation

Round table

Another workshop will be in autumn of 2013 at the Institute of the Agricultural Engineering of the Kassel University in Germany will comprise topics of soil physics and numeric simulations of the water movement in irrigated soils, especially with the innovative subsurface irrigation system.

5   Conclusion

The paper presents a joint research project financed by the Federal Ministry for Education and Research on investigations and development of an auto-regulative subsurface irrigation technique with membrane materials, including qualification measures for the involved scientific staff. Irrigation trials in the partner countries in Africa tested the irrigation system and compared its efficiency and suitability with the common drip irrigation. First results were positive and proof the auto-regulative and water saving potential. The water consumption of the innovative technique was 70 % less (ENSA trial in Algeria) than the compared drip irrigation. The soil surface stayed totally dry so no losses by evaporation could occur. The trials in the partner countries are still running at the moment, further results will be given in future publications.

6   References

Hübener, R. (2006): The Change of the Irrigated agriculture. – Der Tropenlandwirt, supplement No. 63, anniversary publication for Prof. Wolff: Sustainable water use in agriculture: 101-109.

Trough bench subirrigation system for tomato and other vegetables: the OFRALSER project

Manuela Capodilupo and Accursio Venezia

Consiglio per la Ricerca e la Sperimentazione in Agricoltura - Centro di Ricerca per l'Orticoltura / Italian Agriculture Research Council - Vegetable Crops Research Centre Via Cavalleggeri 25, 84098 Pontecagnano (SA) Italy.

Poster. The closed-loop system is among the most environmentally sustainable soilless culture innovations, minimizing water and fertilizer wastes. The nutrient solution is frequently readjusted to plant requirement and disinfected to reduce the risk of disease outbreaks. With trough bench and other subirrigation systems of recirculating nutrient solution, the substrate bottom is in contact with the solution and water and nutrients are carried upward through the root-zone by capillary flow, ending in the substrate top layer, so that ions exceeding plant absorption do not reflow in the recirculating solution, as in drip irrigation, while plant growth is not impaired by the high concentration of salts at the substrate surface. The advantages of the system can be summarized as: 1) stability of the nutrient solution; 2) uniformity of water and nutrient distribution; 3) low incidence of disease; 4) low compaction of the substrate 5) reduction of labour by better use of greenhouse surface and mechanization of many tasks. A decade of research on soilless cultivation of cherry tomato by trough bench irrigation system has been conducted at the Italian Agriculture Research Council - Vegetable Crops Research Centre (CRA-ORT), aimed at an alternative to drip irrigation, which is predominantly used in Italy for fast growing vegetable species. The main factors studied include: tomato cultivar and root stock, concentration of the nutrient solution, substrates, quality of irrigation water, irrigation frequency, watering duration, mulch, frequency of water reintegration. In an on-going project (High - Convenience Fruits And Vegetables: New Technologies For Quality And New Products, OFRALSER) the possibility of prolonging the tomato cycle to more than 30 trusses, of growing other vegetables and of reusing the substrate for further soilless cultivation of tomato or other vegetable species is investigated,

focusing on avoidance of possible autotoxic effects due to root residues and on non-pathogenic micro-organisms in recirculating solutions and substrates, for a possible role in the decomposition of root residues and against soil-borne pathogens.

Keywords: Hydroponics, Solanum lycopersicum, Soilless cultivation, Closed system, Subirrigation, Substrate.

The IRMA project: Efficient Irrigation Management Tools for Agricultural Cultivations and Urban Landscapes

A. Parente[1], Y. L. Tsirogiannis[2], M. Nicolaos[2], B. Pantelis[2], F. Montesano[1]

[1] National Research Council (CNR) - Institute of Sciences of Food Production (ISPA), Italy.
[2] Faculty of Agricultural Technology, Technological Educational Institute of Epirus, (TEI of Epirus), Greece.

Poster. Greece and Italy use about about 70% and 40%, respectively, of the available water resources for irrigation purposes (FAO-AQUASTAT estimations). According to directive 2000/60/EC, action is needed to protect water in qualitative and quantitative terms. Optimization of irrigation management is a key-point for water saving. In fact, this approach has its advantage in the efficiency regarding outcome/cost relationship.

The general objective of IRMA project is to establish a network of knowledge and expertise which will lead to the development of practical irrigation management tools for demand driven capitalization of scientific knowledge and good practices. The partnership is composed by six Greece and Italian partners: Technological Educational Institution of Epirus (TEIEP/Research-Committee, Dept. Floriculture-Landscape Architecture) as lead partner, Decentralised Administration of Epirus–Western Macedonia (ROEDM), Development Enterprise of Achaia- Region of Western Greece (NEA), Region of Puglia (ROP), Istituto Nazionale di Economia Agraria / Bari Branch (INEA) and National Research Council - Institute of Sciences of Food Production (CNR-ISPA).

The project is a part of the European Territorial Cooperation Programme Greece-Italy 2007-2013, priority axis "Strengthening competitiveness and innovation", specific objective "Strengthening interaction between research/innovation institutions, SMEs and public authorities".

In the framework of IRMA project, in accordance with the main objectives to be achieved, the following activities will be carried out: a) development of networking -expertise transfer mechanisms; b) survey regarding the local irrigation practice (legislation, administration, delivery and distribution systems, applied techniques); c) development, applications and evaluations of tools to increase efficiency of agricultural and landscape irrigation (auditing procedures, web information system, creation of knowledge and practical guidance regarding irrigation scheduling, counselling for draught tolerant cultivations, sensors for irrigation management and alternative sources of water) and d) actions for public consciousness building and professional training-certification regarding strategies and methods for efficient irrigation management.

The project is expected to benefit productive agricultural SMEs, agricultural counselling SMEs and local communities by contributing in considerable water and energy savings from improved irrigation efficiency and development of strong cross-border links between agricultural stakeholders, public administration authorities and academic-research institutions in the field of irrigation management.

In the framework of the project 2 open international conferences under the theme: "Efficient Irrigation Management Tools for Agricultural Cultivations and Urban Landscapes" will be organized. One in Patras/Greece and one in Bari/Italy for which cooperation with relevant international organization is expected.

The project has a duration of two years from April 2013.

Evaluation of Subsurface Irrigation Practice in Sub-Sahara Africa

Andrea Dührkoop and Prof. Dr. Oliver Hensel

Kassel University, Department for Agricultural Engineering, Nordbahnhofstr. 1a, 37213 Witzenhausen, Germany.

Abstract. A joint research program coordinated by the Kassel University establishes links between projects dealing with subsurface irrigation (SI) methods in Ivory Coast, Algeria, Kenya, Turkey and Namibia. Within the framework each project will conduct an analysis of SI methods in field trials under local conditions in terms of water productivity, use of low quality water, salinity effects in the soil and socio economic aspects. One main objective of this project is to summarize the results from different SI projects into guidelines for stakeholders, farmers, water managers, water authorities and ministries.

Keywords: subsurface irrigation, water efficiency, porous pipe, pitcher irrigation, subsurface drip irrigation.

1 Why this project

The implementation of subsurface irrigation (SI) techniques aims on increasing water productivity – gaining more yields and value from water is an effective means of intensifying agricultural production and reducing environmental degradation. The reduction of water consumption for irrigation purposes leads to a significant improvement in the security of the inhabitants of agriculturally dominated regions with limited water resources, both in terms of water supply (primary effect) as well as the food supply (secondary effect).

2 Support program for the project

The project will be conducted in the frame of the European Research Area for Agricultural Research for Development program (ERA-ARD II) funded by the European Commission's Framework 7 Programme. ERA-ARD II will address these issues through improving coordination and collaboration between national research programmes. The ERA-ARD project is a partnership of 17 organisations involved in funding agricultural research for development in 15 European countries. In Germany the Ministry in charge is the Federal Agency Agriculture and Food (BLE) providing financial support for the presented project.

3 Project description

Africa currently lags behind all other regions in terms of farm productivity levels, with depressed crop and livestock yields. Its agricultural sector is confronted with numerous constraints, such as soil erosion and degradation, increasing salinity, excessive tapping of groundwater and persistent droughts (IFAD 2010[1]). A major part of these problems can be related to poor irrigation technology.

The commonly used, accepted and low cost irrigation technique is the traditional surface irrigation (flooding) which is water wasting and not environmentally friendly. A more efficient and environmental friendly technique is the subsurface irrigation (SI). The main advantages of this technique are the dry soil surface (evaporation is minimized), dry plant foliage (prevents plant diseases) and the effective use of water due to the direct application in the plants' root zone. These known system advantages make the SI techniques very favourable with respect to the expected impacts of climate change like water scarcity, rising temperature and thus higher evaporation rates.

The project coordinated by the Kassel University will establish links between projects dealing with SI methods in Ivory Coast, Algeria, Kenya and

---

1   IFAD – International Fund for Agricultural Development 2010:
     http://www.ruralpovertyportal.org/web/guest/country/approaches

Namibia. Within the framework each project will conduct an analysis of SI methods in field trials under local conditions in terms of water productivity, use of low quality water, salinity effects in the soil and socio economic aspects.

Nevertheless, as SI is a rather new approach, different approaches exist that are based on different technological concepts. These concepts are analysed and evaluated in several independent research activities, which will be bundled into the above mentioned research project:

- Subsurface drip irrigation (SDI) (buried dripper lines), as derived from conventional drip irrigation (currently investigated in the frame of the Association for the Development of Intensive Food Crops in Côte d'Ivoire (ADCVI) as well as in the Small scale Producers of vegetables in the Savanna region (= Petits Producteurs Maraîchers dans les régions des Savanes, PPMS) which is a development support program dealing with food crop chains in the savanna regions (northern part of Côte d'Ivoire)
- Porous hose made of used tires granular (investigated in the frame of an IWRM-project in Namibia and Côte d'Ivoire)
- ARSIT (Auto regulative Subsurface Irrigation technology, developed by the chairing institute for Agricultural Engineering in the frame project financed by the German Minstry of Education and Research (BMBF).

The project aims at aligning these projects and the involved capacity-building activities, in order to achieve results that will finally allow comparing and evaluating existing SI technologies in order to summarize the results into guidelines for stakeholders, small scale farmers, water managers, water authorities and ministries.

The main part of the necessary field trials, laboratory testing and socio-economic investigations will be performed in the frame of existing research projects.

The project is divided into 8 work packages (WP) with certain tasks:

WP1	Management and Coordination
WP2	Collection of local knowledge
WP3	Socio-economic assessment for the implementation of the SI techniques
WP4	Environmental impacts of the SI when using low quality or treated waste water
WP5	Installation technology
WP6	Summarize the achieved results in guidelines for an adoption by the small scale farmers
WP7	Numeric modelling of the water and solute transport in each subproject for evaluating water productivity and environmental risks
WP8	Capacity building

The project will induce important progress in comprehensive knowledge on SI-technology. It will allow maximizing the quality and skill diversity of the expertise base by strengthening research and training links between African institutions, between African institutions and northern partners, and between francophone and anglophone countries (north-south and south-south cooperation). The project results will allow an optimized choice and application of appropriate technology with respect to specific local conditions. It will thus contribute to an improvement of the environmental and economic situation of farmers in the affected countries and beyond.

4 Conclusion

The authors present a joint research project which aims at evaluating different subsurface irrigation techniques and increasing their implementation. The research project will establish links between existing projects which are dealing with SI methods. In each of these projects field trials with different subsurface irrigation will be conducted and evaluated with respect to different boundary conditions (climate, soil, water quality etc.). On the scientific level, the project will provide reliable data that will allow evaluating different SI-technologies. The results will finally allow

establishing guidelines for the use of subsurface irrigation as an efficient and environmental friendly irrigation technique. The guidelines will be provided to stakeholders, farmers, water managers, water authorities and ministries.

Competence Modelling for the Agricultural Sector: The AGRICOM Project

Cornelia Helmstedt, Xenia Rodriguez and Christian M. Stracke

Information Systems for Production and Operations Management
University of Duisburg-Essen, Germany.

Abstract. This article presents the AGRICOM Project. The main goal of the AGRICOM project is to establish the first Competence Model for the Agricultural Sector (ACM) in order to strengthen the transparency and comparability of VET opportunities at a European level. AGRICOM stands for Transfer of the Water Competences Model to AGRIcultural COMpetences and is supported and funded by the European Commission.

Keywords: AGRICOM, Competence Modelling, competence, agriculture

1 Introduction

Due to the shortage of the valuable resource water and the consequences of climate change, numerous European policies have been developed and adopted for the protection and sustainable utilisation of water also with impact on vocational education and training (VET). Economic factors like privatisation and increasing cost pressure in water management are increasing these VET needs and thus have led to an increased demand for modularised and tailor-made VET opportunities.

2 The AGRICOM Project

One project to address this demand and to make VET more responsive to labour market needs in the agricultural sector is the AGRICOM project.

AGRICOM aims to improve the identification and anticipation of skill and competence needs and their integration in VET provisions. In particular AGRICOM supports the implementation of the "New Skills for New Jobs" (http://www.na-bibb.de/index.php?id=1518) strategy by taking into account challenges, such as environmental and demographic changes, and related growing job needs in the agricultural sector. "An agenda for new skills and jobs" to modernise labour markets and empower people by developing their skills throughout the lifecycle is one of the seven flagship initiatives of the strategic framework for the European Digital Agenda (DAE) and the core focus of the long-term framework for European cooperation in education and training (ET 2020). The on-going development and management of knowledge, skills and competences at the individual and organisational level is therefore set as a clear priority and challenge for the EU and its Member States, for vocational education and training, for companies, employees, and learners.

To facilitate a better relationship between the professional and work life, skills, competence needs, and their implication in vocational education and training have to be identified and anticipated. Competence Modelling is becoming more and more crucial not only for business success but also for the European society and citizens. It will become an integrated part of both organizational and individual lifelong learning strategy. The current challenges are to address learning outcomes and competences and their modelling within the given frameworks of national and regional education systems and regulations.

Thanks   to   the   Bologna   Declaration (http://ec.europa.eu/education/policies/educ/bologna/bologna.pdf), learning outcomes and competences are already integrated in higher education, but are just now becoming important for VET. Therefore European projects in the VET field are most important for the promotion and sustainable establishment of Competence Modelling.

AGRICOM capitalises and transfers the results of prior European LdV-projects: on one hand, the Competence Model developed for water resources management in the WACOM Project (www.wacom-project.eu) as the core input, and on the other hand, the instruments for the integration of competences in the EU policies EQF and Europass developed in the

framework of the e-competences tools eCOTOOL project (www.ecompetence.eu).

The WACOM Competence Model transferred the European Qualification Framework (EQF) and the German Reference Model for the Competence Modelling PAS 1093 into the water sector and its Vocational Education and Training (VET) throughout all of Europe. The overall objective of eCOTOOL was to improve the development, exchange, and maintenance of VET certificates as well as their accessibility and transparency by harmonising Europass with other European instruments (such as EQF, ECVET) and e-competences to provide a harmonised and standardised structure for the definition and description of competences and skills.

The European AGRICOM project advances the development of competence modelling in Europe through its application to one of the most important European vocational areas – the agricultural sector, and in particular the fields related with the unitization of the water resources in irrigation, hydroponics and other agricultural fields.

The main product of AGRICOM is the AGRICOM Competence Model (ACM), which provides all elaborated competences for the agricultural sector, templates for competence descriptions, job profiles and competence profiles, and use examples of a single competences or job profiles and trainings opportunities.

During the first project year, the AGRICOM Consortium concentrated on needs analysis, covering national workshops, desk research, and an online survey, as well as the transfer and adaptation of the ACM, whereas in the second year the focus has been on the pilot testing and optimization of the ACM, as well as the dissemination and exploitation of all project results.

The intensive and comprehensive pilot-testing phase took place over six months in four countries and ensured a long period of evaluation, validation, and optimization of the ACM. This supported further outcomes as well as many implementations and applications throughout Europe.

3 Conclusion

The AGRICOM Project established the first Competence Model for the Agricultural Sector, in the form of the ACM, in order to strengthen the transparency and comparability of VET opportunities on the European level by transferring an existing Competence Model from the water sector to the agricultural sector.

The AGRICOM project has thus built a basis for harmonization on a European scale through the combination of leading European instruments and aforementioned frameworks. The ACM facilitates the continuous development and management of knowledge, skills, and competences at both individual and organizational levels. This is clearly set as a priority and task for the EU and its member states, as well as education and training providers, companies, workers, and learners.

Through AGRICOM, a competence and qualification model for the most important field of the agricultural sector has been established exemplarily for hydroponics and irrigation of crops that can be transferred easily into other fields in the agricultural sector (and also into other sectors). Modular VET opportunities and products in the field of hydroponics and irrigation of crops, as well as other agricultural fields, can be defined precisely corresponding to the addressed competences and qualifications, becoming transparent and easy to select as they accurately fit the needs of their users.

In conclusion, AGRICOM has provided an effective Competence Model through the consolidation and technical representation of competence definitions, thus enhancing the transparency and comparability of training opportunities and job profile descriptions at a European level.

8 References

ET 2020 (2009) = Education and Training 2020. Brussels: European Commission [online available on: http://europa.eu/legislation_summaries/education_training_youth/general _framework/ef0016_en.htm].

Stracke, C. M. (2011a). Competence Modelling for Innovations and Quality Development in E-Learning: Towards learning outcome orientation by competence models In Proceedings of World Conference on Educational Multimedia, Hypermedia and Telecommunication 2011 [= EdMedia 2011] (p. 1885-1894). [also online available on: http://www.qed-info.de/downloads].

Stracke, C. M. (2011b). Competence Modelling, Competence Models and Competence Development – An Overview In STRACKE, C. M. (ed.). Competence Modelling for Vocational Education and Training. Innovations for Learning and Development (p. 11-33). [also online available on: http://www.learning-innovations.eu/].

Methodology for Transfer of the Water Competence Model into the Agricultural and Horticultural Sector

Henrike Perner

Leibniz-Institut of Vegetable and Ornamental crops, Theodor-Echtermeyer-Weg 1, 14979 Großbeeren, Germany.

Abstract. This article describes a method to transfer the water competence model (WCM) into the agricultural and horticultural sector. Both models were developed in projects on lifelong learning to increase the transparency of educational programs and the mobilization between the European countries.

Italy, Germany, Greece, and the Netherlands participated to collect data in terms of competences, vocational and educational training, job profiles and competence modelling in the special field of irrigation and hydroponics. The described agricultural competence model (ACM) consists of four main components: (a) stakeholders, (b) competences, (c) template for job profiles, and (d) template for use case scenarios and learning outcomes. The templates are complemented by Entity-Relationship-Diagrams.

These templates can be used e.g. to create individual job profiles or to compare the educational levels and teaching content of training unit between countries.

Keywords: water competence model (WCM), agricultural competence model (ACM), hydroponics, irrigation, transfer, competence, job profile, training units, use case scenarios, and learning outcomes.

1 Introduction

The Water Competence Model (WCM) and the Agricultural Competence Model (ACM) were developed in the projects WACOM and AGRICOM, respectively. They support the overall goal of the European Union to increase the transparency of educational programs and the mobilization between the European countries.

In general, competence modelling is based on the development of a database relational model using templates for description of the data (Ghanei, 1999). These templates are complemented by diagrams to describe the relation of the different data (Entity-Relationship-Diagram). In case of two models describing vocational and educational training (VET), the educational quality standards of the EU are important. Therefore, the indication of the qualification levels within the WCM and the ACM are based on the COUNCIL DECISION of 16 July 1986 on the comparability of vocational training qualifications between the Member States of the European Community (85/368/EEC) and are in line with the RECOMMENDATION OF THE EUROPEAN PARLIAMENT AND OF THE COUNCIL of 23 April 2008 on the establishment of the European Qualifications Framework for lifelong learning and the German specification PAS 1093 on competence modeling.

For the methodological transfer of the WCM into the agricultural and horticultural sector, especially into the area of irrigation and hydroponics, the templates of the WCM were investigated, described and adjusted to the ACM. Results of collected data within the fields of agriculture and horticulture of the four European countries Italy, Germany, Greece and the Netherlands were analysed and integrated into the templates of the ACM model. Data were collected on a) competences, competence models and needs for a competence model in the different fields of agriculture and horticulture, b) stakeholders that might be interested in a competence model, c) information about quality standards and general knowledge on EQF in the partner countries, and d) different VET models already used in the partner countries.

In this text the WCM and its methodological transfer to the ACM are described in short.

2 Water Competence Model

Several steps have been taken to develop the WCM. First the water sector was analysed and a pool of competences collected. These competences were grouped and defined (competence = skills, knowledge, ability, qualification level). From this competence pool several templates were generated, including job profile descriptions, trainee profiles, training opportunities and certificate supplements. Core elements of the WCM instrument to create training curricula were the competence catalogue and the competence profiles (Fig. 1). A competence profile describes a pool of competences of a certain unit, e.g. all jobs performed on a certain farm. The competence model defined competences that each training unit contained.

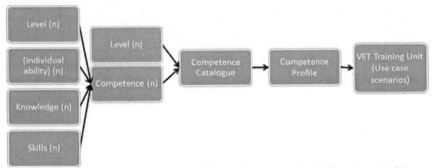

Figure 1: Example for a competence modeling pathway towards a VET training unit of the WACOM model. By clustering competences into units, individual training curricula and learning opportunities can be created. (n) Stands for numerous possibilities.

The two main components of the WCM were competences and qualification levels (WACOM, 2011). The competences were divided into *key competences, individual competences*, and *water-specific competences*. The *key competences* were based on the European Policy key competences of the European Commission. The *individual competences* and *water-specific competences* were generated from research done in the water sector. Additionally, five qualification levels described the standard for learning outcomes, following the recommendations of PAS 1093 and the European Qualification Framework (EQF). Four templates have been developed to describe different situations in the area of VET: WCM templates for

competence description, for job profiles, for competence profiles and for use case scenarios.

To transfer the WCM into the agricultural and horticultural sector it was necessary to investigate first the state of art in the agricultural sector. From the background of this analysis of the agricultural and horticultural sector the developed templates of the WCM were adapted to the needs of the ACM. Additionally, learning outcomes were described as a valuable basis for the development of the ACM.

3 Agricultural Competence Model

The first step taken was the identification of possible stakeholders that might use the ACM in future. They were chosen on the basis of data collected in workshops, personal and online surveys, interviews, and personal experience of the project partners. The following groups were identified: agronomists related to hydroponics or irrigation in the private and public sector, individual and associated growers, private and public authorities related to VET, learners, government, local and regional authorithies, VET trainer, hydroponical start-ups and investors, technology/service provider and supplier, and environmental bodies.

Competence

A competence was defined as a combination of skills, knowledge and the individual attitude and ability to perform a job. In personal interviews and during national consultation meetings competences for professionals in irrigation and hydroponics have been gathered. Three different categories of competences have been described: *key competences, managerial or individual competences* and *professional or agricultural/horticultural competences.* The professional competences have been divided into two groups, plant and water (Table 1).

A. Key competences
Social competence
Communication
Professional practitioner
IT-Competence and building of networks and online communities

B. Managerial or individual competences
Management, e.g.: business administrative, human resource and resource management, maintain profitability, production plan, quality management
Safety at work
Self development
Product safety
Legal basic knowledge

C. Professional or agricultural/horticultural competences	
Plant nutrition	
Plant production	PLANT
Plant hygiene and protection also in water management	
Soil and substrate science	
Water resources, qualiy and storage	
Sustainable water management	WATER
Process control	
Selection of the appropriate system and the supporting technical equipment	
Steering of hydroponic and irrigation technique (climate control or meteorology)	
Consideration of nutrient evaluation for optimisation of irrigation	

Table 1: Selection of competences for the agricultural and horticultural sector that are useful for the maintenance of irrigation and hydroponic systems.

Job profile

The template for job profiles (Table 2) used for the ACM includes definition of the job profile, target group, the context in which this job is performed and level. On the basis of the analysis of the agricultural and horticultural sector the context and the general qualification level were added to the WCM template of the job profile.

ACM Job profile	"Title"	
Definition of job profile		
Target group of job profile		
Context in which job is performed		
Level		
Competences contained in and required for job profile (comp. catalogue)		
Name of competence	Definition	Required level

Table 2: ACM Template for a job profile

The templates for competence profiles, use case scenarios and the use case diagram were transfered from WCM with minor changes.

Use case scenarios, diagrams, and learning outcomes

In competence modelling the search path towards an individual training unit/module must be described in a process oriented way (Ghanei, 1999). In order to consider all stakeholder perspectives, use case scenarios were developed. The perspective was not only important, but also the communication pathway to describe the relation of the different stakeholders within one use case scenario (Ghanei, 1999). This was done by adding use case diagrams. Depending on which stakeholder wanted to make use of a training unit, different use case scenarios were developed. One use case scenario with a matching diagram was a system to determine the needs of one particular stakeholder for specific competences from one particular

perspective (Jacobsen, 2011). The sum of use cases made clear what a system was going to do and what it was not going to do.

Use cases have been utilized to develop software systems since their initial introduction at "Conference on Object-Oriented Programming Systems, Languages, and Applications in 1987" (Meyrowitz, 1987). Over the years they have become the foundation for many different methods and an integral part of the Unified Modeling Language (Jacobsen, 2011). They are used in many different contexts and environments, and by many different types of teams. For example use cases can be beneficial for both small agile development teams producing user-intensive applications and large projects producing complex systems of interconnected systems, such as enterprise systems, product lines, and systems in the cloud.

The ACM use case template includes a title, summary, stakeholder, stakeholders' goal, and need for individual learning pathways, competence profiles, job profiles, a competence catalogue, use case procedure, and unit. It describes the path to achieve a certain goal from different stakeholder perspective. If he or she is an employee who wants to achieve certain competences, it describes the way from his or her employer to the VET provider. On the other hand it could also show how a VET provider offers particular competences to an employee out of his or her pool of training units and competences.

The diagram below shows the relationship between the different stakeholders (Figure 2). It clarifies the correspondence paths between the different stakeholders. This facilitates the creation of a database which can be saved and processed in a computer-based system.

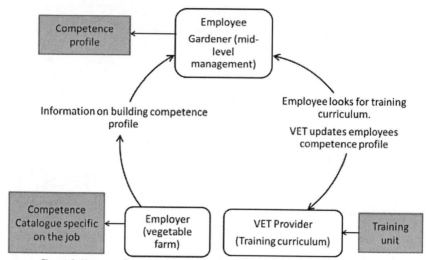

Figure 2: Use case diagram, showing the interactions paths between the different stakeholders.

Using the competences for the description of job profiles and use case scenarios revealed that the same competences were used for different levels of education. It was assumed that the same terminology could be utilized with increasing qualification levels, but at the same time a broader knowledge was hiding behind the competences. To describe these differences, competences also have been described in terms of general learning outcomes. A template has been developed additional to the templates used in the WCM to describe learning outcomes.

The implementation of learning outcomes for the description of competences is one goal of the European Union. They hope to increase permeability and improve the dialogue between the different education institutions by implementing national qualifications frameworks (NQF) (PLA, 2009). Mapping qualifications against the same set of descriptors makes it apparent where two (or more) qualifications lead to comparable learning outcomes and what the learners additionally need to achieve. A guideline for the formulation of learning outcomes (BMBF, 2012) describes learning outcomes as statements of what a learner knows, understands and is able to do on completion of a learning process. Learning outcomes are defined in terms of knowledge, skills and competence.

The ACM template for describing learning outcomes for different use case scenarios includes title, competence catalogue and learning outcomes (Table 3).

Titel (One-sentence title of the Use Case)	
Competence Catalogue	**Learning outcomes:** He/she is able to…
Competence (n)	Knowledge: …. Skills: … Competences: …

Table 3: General template for learning outcomes for a use case scenario. (n) Stands for numerous possibilities.

One of the main parts of the ACM was the interaction of different competences forming a certain training unit/module. A competence, including skills, knowledge, and the individual attitude and ability to perform a job, were combined into an activity area (Figure 3). The competence catalogue, in combination with the quality level and the context in which a certain job is performed, can be used to describe specific job profiles, competence profiles, use case scenarios, and VET training units/modules. Additionally, learning outcomes have been included in the ACM to improve the comparability of competences on different qualification levels.

Figure 2: Example for a competence modelling pathway towards a VET training unit of the AGRICOM model. A detailed description of the competencies forms the competence catalogue. Out of this competence catalogue, job profiles, including competence profiles, can be created. Using the competence profiles a use case scenario with specific training unit can be generated. (n) Stands for numerous possibilities.

The usage of the ACM can have various forms. One possibility would be to describe all job profiles in a specific company to support the search for adequate human resources. Another possible application of the ACM would be its integration into an online tool. For example, on a web portal an employer, searching for new staff, can choose the adequate competences out of a list of competences to write a plain job application. On the other hand, an employee who wants to improve his professional skills might use this tool to reveal missing competences in his job profile, a so called gap analysis. This gap analysis of course can just as well be used by an employer.

4 Conclusion

In conclusion it can be said that the transfer and adaptation of the WCM into the agricultural and horticultural sector in form of the ACM should be possible with the described metholodology. Further work will be needed to standardise competence definitions, clearly describe working places, job profiles, and training solutions.

In the end, the user might use the ACM and its tools in several contexts, e.g. for the development of job advertisements, description of job profiles, or for writing curricula. In this way, the competence model would support the need in agriculture and horticulture for a better match between the labour market and VET.

5 References

BMBF, 2012, Geographical Mobility in Vocational Education and Training: Guidelines for describing units of learning outcomes, Retrieved July 22, 2013 from   http://www.ecvet-info.de/ media/Guidelines for describing units of learning outcomes.pdf

Ghanei A. (1999): *Datenorientierte Sicht des Unternehmens, in Systemanalyse im Unternehmen*, Krallmann H., Frank H., and Gronau N. (Edt.): 3rd Edition, Oldenburg Verlag München Wien, pp 144-169

Jacobson, I., Spence I., Bittner K. (2011) *Use Case 2.0: The Guide to Succeeding with Use Cases*, IJI SA

Meyrowitz, N.K. (Ed.) (1987). *Conference on Object-Oriented Programming Systems, Languages, and Applications* (OOPSLA'87), Orlando, Florida, USA, October 4-8 1987, Proceedings.

PLA (2009). *PLA National Qualifications Frameworks - bridges between HE & VET in terms of learning outcomes* –Summary Report 15-16 June 2009, Berlin, Germany. Retrieved July 22, 2013 from http://www.kslll.net/Documents/Recognition%20of%20learning%20outcom es_Report_on_Berlin_PLA.pdf

WACOM (2011). *The Water competence model*. Retrieved July 22, 2013 from http://www.wacom-project.eu/

A Quality Management System for Hosting AGRICOM Representations

Andreas Drakos, Charalampos Thanopoulos, Yannis Psochios

Agro-Know Technologies, Athens, Greece.

Abstract. The AGRICOM project establishes the first competence model for the agricultural sector in order to strengthen the transparency and comparability of training opportunities at European level. One of the main outputs of the AGRICOM initiative is the creation of a web portal and of a repository for storing and hosting Vocational Education and Training (VET) elements based on the AGRICOM Competence Model (ACM). To ensure the quality of the elements hosted in the AGRICOM portal, a specific quality management system has been implemented. The system forces uploaded content through a specific workflow that requires user interaction and validation before publishing; hence by involving users it certifies that specific quality criteria are met.

Keywords: Agriculture, AGRICOM, Quality Management System, Competence Model, VET

1 Introduction

AGRICOM project (http://www.agricultural-competences.eu), Transfer of Water Competence Model to AGRIcultural COMpetences, is a European Project in the context of Lifelong Learning Program, which intends to support the professionals and learners in the agricultural sector with the identification of required competences for the successful work performance in specific agricultural fields. The identification of agricultural competences can serve two different aspects. On one hand competences are defined in a way to describe the learning outcomes of learning courses [0], [0] and on

the other hand a list of essential and desired competences sets the outline of the job profile of the employees in specific working framework [0], [0], as farmers or workers with specific duties in the production chain of a farm (e.g. harvesting techniques, taking care a vegetables garden or experienced agronomists in handling pests and diseases management in terms of prevention and control them).

The AGRICOM Competence Model (ACM), the main outcome of AGRICOM, provides a standardised and harmonised competence model for the agricultural sector [0]. Apart from the ACM model, the AGRICOM project also aims in creating an online tool – a web portal – for the representation of the ACM. The AGRCIOM web portal provides a user-friendly tool with a quality control system for the enhancement of the ACM dissemination in a greater amount of experts and interested users from the whole Europe. This will support the further evaluation of the ACM by experts from several fields of the agricultural sector and it will give the opportunity to professionals, employees and employers to test the structured Competence Model in real cases of the job professions and training opportunities in the agricultural sector.

In this paper we present the AGRICOM Web Portal as the mean to host and represent ACM elements and the quality control system that is implemented to validate the numerous uploaded elements. The rest of the paper is divided as follows: in the next chapter we present the different categories of elements that can be stored and hosted in the AGRICOM Web Portal, the following chapter presents the general architecture of the portal and the different user roles that are present all of which are necessary to explain in the next chapter the quality control system used in AGRICOM and the final chapter that concludes this paper.

2 Online Representation of the ACM

The main goal of the AGRICOM Web Portal[2] is to support the development of the establishment of the first Competence Model for the Agricultural Sector (ACM) in order to strengthen the transparency and comparability of VET opportunities at European level. The portal lists, collects, describes and categorizes vocational training content based on the competence building block and represents it in a user-friendly environment.

The portal is the online tool that implements the AGRICOM Competence Model and allows the creation and population of competence profiles for different cases in the agricultural sector. It is used to describe a number of learning elements, the description of training opportunities and the related certificates and tasks and responsibilities for defined working places. These elements have been further categorised, depending on their purpose as follows:

- Competence Elements (CEs), descriptions of competences in the agricultural sector, including skills and knowledge. Competences are a building and supporting block for all other Elements.
- Job Profile Elements (JPEs), represent professions in the agricultural sector, including workplaces descriptions and are registered according to the category - classification to which they belong e.g. Educational support activities.
- Training Opportunity Elements (TOEs), descriptions of training opportunities related to the agricultural sector, registered based on their classification and includes both required and targeted CEs while they can be connected to specific JPEs and targeted Certificates.
- Certificate Description Elements (CDEs), represent certificates that can be attained through specific TOEs to certify that a person has specific CEs

The intention of the categorization is from one point of view to help Content Providers upload and annotate their material, while at the same

---

[2] http://portal.agriculture-competence.eu/

time help all users (e.g. professionals, training providers) in their search for particular material. All elements that are described in the portal follow the AGRICOM Competence Model (ACM).

3 General Architecture of the Portal

Figure 1 illustrates the main architecture of the AGRICOM Web Portal. As it can be seen in the figure, in the portal different kind of users have been defined based on the services provided to each type. Each type can access different functionalities of the site and can interact with the different available databases. The different types of users and the services that are provided to them are the following:

- **Visitor**, can use the Public Services of the Web Portal such as browsing or searching for CEs, JPEs, TOEs and CDEs.
- **Content Provider,** can upload CEs, JPEs, TOEs and CDEs and their corresponding Metadata.
- **Content Annotator,** can further annotate uploaded CEs, JPEs, TOEs and CDEs.
- **Content Translator,** responsible for translating the uploaded elements in other languages.
- **Content Validator,** responsible for validating a CE, JPE, TOE and CDE, before the element is made available to the public.
- **Administrator,** perform all the administrative functions related to Users, These include Viewing / Deleting / Deactivating / Validating CEs, JPEs, TOEs and CDEs, Accepting or Declining Requests for Registration from Users, Viewing / Activating / Deactivating Users etc.
- **External Tools** (e.g. Harvesters) interfacing with the Repository and consuming content.

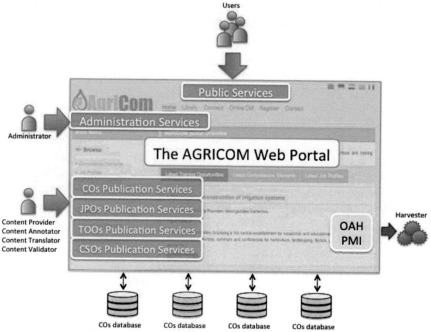

Figure 1: Overall Architecture of the AGRICOM Web Portal

Another main characteristic of the AGRICOM Web Portal is multilinguality. Providing a service in different languages is a very important factor to facilitate users in searching and browsing content throughout the portal. In the AGRICOM portal all pages are available in the following five languages: Dutch, English, German, Greek and Italian.

4 Quality Management System

In order to ensure the validity and the quality of information uploaded to the AGRICOM Web Portal a specific process regarding the information flow is implemented. Releasing everything to the public as soon as it is uploaded would diminish the quality of information provided by the web portal. Thus, although the process requires the actors to engage higher effort, it ensures as much as possible, the quality of the information.

AGRICOM uses a specific Quality Management System in order for the uploaded content to be validated before being published. The validation of the content refers to the following two criteria: the content must have full metadata descriptions and be translated in the different languages that are supported by the portal. The generic version of quality management system is consisted of the following:

- Only registered providers can upload content
- Annotators are available to inspect and annotate content when needed
- All content must be translated in all the available languages
- Before publication, the content must be validated

The above process ensures the quality of the content that is being published in the AGRICOM web portal.

The implementation of the quality management system in the AGRICOM portal leads to the following workflow for uploading new content:

- Step 1: a Content Provider uploads an Element (competence, job profile, training opportunity or certificate) to the AGRICOM Web Portal in his own language and in the English language.
- Step 2: the administrator assigns to a specific Annotator the Element for annotation in the language of the Element.
- Step 3: the Annotator annotates the Element in his own language and in the English language.
- Step 4: the administrator assigns the Element to one Translator for each language in order to provide information in all languages of the Web Portal.
- Step 5: each Translator submits the translated version of the Element.
- Step 6: the Administrator assigns the Element for validation to a Validator.
- Step 7: as soon as the Validator reviews all the information the Element, it is published and is now available for the portal visitors and for harvesting processes.

A schematic version of the steps described above is shown in the figure bellow.

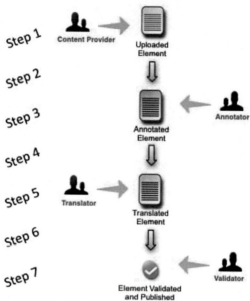

Figure 2: AGRICOM Workflow, implementing the quality management system

The AGRICOM Workflow for quality management can also be explained through the following example. John is a Human Resources (HR) manager working in a Greek company related to the agricultural sector. John has learned about the AGRICOM web portal and believes that the portal is a nice way to describe job profiles related to his company and connect them with required competences. John, following the information available at the portal, contacts the AGRICOM administrator to register as a content provider user. When his account is enabled he uses the portal to upload his first job profile with the description of "Cultivation Specialist".

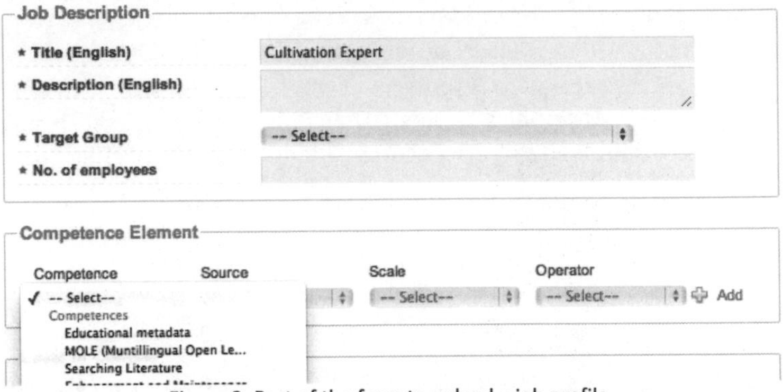

Figure 3: Part of the form to upload a job profile

To do so, John is redirected to a specific page where a form is available to describe the job profile and connect it with a number of different competences as shown in Figure 3. John needs to fill the form with all the necessary information regarding this specific Job, to add the title and the description and finally from a drop down menu he can access the AGRICOM list of competences to select and relate with this specific job.

As soon as John uploads the job profile, it enters the AGRICOM workflow. The administrator first assigns to a user from the pool of Annotators in Greece, for this job profile to be annotated, i.e. check if all the fields are correctly filled and make any corrections necessary. When the job profile is successfully annotated, the administrator is responsible to assign translators so it is translated to all the supported languages. After the translation, as a final step, the administrator assigns this job profile to a validator who will check if all the process is successfully finished and will finally publish the job profile.

5 Conclusion

The main idea of the design and development of an online representation of the ACM is to provide a user-friendly environment where the description of job profiles, training opportunities and certificates is mapped with the needed and required competences and skills. The AGRICOM web portal is open to users across the world and in order to ensure the quality of the

uploaded content it implements a specific quality control system. This system is incubated in the upload process of the AGRICOM web portal, the AGRICOM workflow, and requires user interaction in order to annotate, translate and validate each element that is being uploaded. Through this process all the uploaded content is validated and a specific quality can always be met.

6 Acknowledgements

This work is funded with the support by European Commission, and more specifically the project No DE/11/LLP-LdV/TOI/147 458 "AGRICOM: Transfer of Water Competence Model to AGRIcultural COMpetences)" (http://www.agricultural- competence.eu) of the Lifelong Learning Programme (LLP). This publication reflects the views only of the authors, and the Commission cannot be held responsible for any use which may be made of the Information contained therein. The authors would like to thank all the consortium partners for their contribution in the design and realization of the requirements analysis.

7 References

Stracke, M.C.: Competences and Skills in the Digital Age: Competence Development, Modelling, and Standards for Human Resources Development. Proceedings of the 5th International Conference, MTSR 2011, Izmir, Turkey, October 2011, ISSN: 1865-0929, pp. 34-46 (2011).

Thanopoulos C., Protonotarios, V. and Stoitsis, G.: Online Web portal of competence-based training opportunities for Organic Agriculture, Agris on-line Papers in Economics and Informatics, Vol. IV (1): 67-86 (2012).

Thanopoulos, C., Manouselis, N., Kastrantas, K. and Psochios, Y.: Design and Development of the ICT Tools for the Online Dissemination of the WACOM Competence Model (WCM). Proceedings of the 4th European Conference: "Innovations in the Environmental Sector" Competence

Models, E-Learning, Social Communities, Brussels, Belgium, September, 21, pp. 82-97 (2011).

Thanopoulos, C., Drakos, A. and Stracke, M.C.: Application Profiling for Agricultural Communities: Transferring the Water Application Profile to the Agricultural Sector. To be presented in 7th Metadata and Semantics Research Conference, Thessaloniki, Greece, 19-22 November (2013).

Author Index

Bencheikh, Abdelaali
Technical and Scientific Research Center on Arid Regions (CRSTRA), Biskra, Algeria
bencheikh1400@gmail.com

Borgognone, Daniela
Dipartimento di scienze e tecnologie per l'Agricoltura, le Foreste, la Natura e l'Energia, Università della Tuscia, Viterbo, Italy
Borgognone@unitus.it

Capodilupo, Manuela
Consiglio per la Ricerca e la Sperimentazione in Agricoltura - Centro di Ricerca per l'Orticoltura / Italian Agriculture Research Council - Vegetable Crops Research Centre, Pontecagnano (SA) Italy
manuela.capodilupo@unina.it

Manuela is working as a research fellow at CRA-ORT (Agriculture Research Council-Vegetable Crops Research Centre) since October 2012.
In March 2007, she was awarded a degree in Agricultural Sciences and Technologies from the University of Naples "Federico II" about "Antimicrobial activity of saponins of Allium minutiflorum". In March 2011, she was awarded her PhD from the same university in Agrobiology and Agrochemistry; title *"Effect of microbial diversity on several ecosystem functions"*.

Cardarelli, Mariateresa
Consiglio per la Ricerca e la sperimentazione in Agricoltura – Centro di Ricerca per lo studio delle relazioni tra Pianta e Suolo, Roma, Italy
mteresa.cardarelli@entecra.it

Colla, Giuseppe
Consiglio per la Ricerca e la sperimentazione in Agricoltura – Centro di Ricerca per lo studio delle relazioni tra Pianta e Suolo, Roma, Italy
giucolla@unitus.it

Djoudi, Madjed A.
Technical and Scientific Research Center on Arid Regions (CRSTRA), Biskra, Algeria
madjeddjoudi@yahoo.fr

Drakos, Andreas
Agro-Know Technologies, Athens, Greece
drakos@agroknow.gr

Andreas received a B.Sc. degree in telecommunications (2006) and a Ph.D. in the field of optical networks (2010) both from the Department of Informatics and Telecommunications of the University of Peloponnese. In 2007 he joined the Optical Networking Group where he worked as a research associate until 2012. He has participated in a number of research projects, EU and Greek national funded, on the field of optical networks.
Andreas joined Agro-Know Technologies in 2013 as a Project Manager. He is responsible for the planning and successful execution of research projects and he is involved in deliverable preparation and project dissemination.

Dührkoop, Andrea
Kassel   University,   Department   for   Agricultural
Engineering, Witzenhausen, Germany
andrea.duehrkoop@uni-kassel.de

Andrea Dührkoop is a PhD student at the Agricultural Engineering Institute in the Tropics and Subtropics at the Kassel University located in Witzenhausen at the Faculty of Organic Agricultural Science. She coordinates research projects focused on sub-surface irrigation which are conducted in close cooperation with Universities, research institutes and industry companies from Africa, Turkey and Germany. Within this network innovative and efficient irrigation technologies will be developped and tested with regard to users' need in terms of manageability, economics and social aspects.

Fiorillo, Antonio
Dipartimento di scienze e tecnologie per l'Agricoltura, le Foreste, la Natura e l'Energia, Università della Tuscia, Viterbo, Italy
fiorillo@unitus.it

Hartani, Tarik
National   Advanced   School   of   Agronomy   (ENSA), Institute for Water in Agriculture, Algeria
rik_hartani@yahoo.fr

Helmstedt, Cornelia
University of Duisburg-Essen, Campus Essen
Information Systems for Production and Operations
Management
cornelia.helmstedt@icb.uni-due.de

Cornelia Helmstedt is Scientific Researcher at the University of Duisburg-Essen (UDE). She studied Educational Sciences and Roman Linguistics at the Leibniz University in Hannover. Since 2010 she is working at UDE in the Department "Information Systems for Production and Operation Management" (WIP). She has got a strong background and expertise in evaluation and in leading, coordinating and contributing to research projects (SEVAQ+, CerOrganic, Concede, OrganicEdunet, COMPATeGov, OERtest; VOA3R).
Currently she is responsible for the research activities in the AGRICOM and Inspiring Science projects.

Hensel, Oliver
Kassel   University,   Department   for   Agricultural
Engineering, Witzenhausen, Germany
agrartechnik@uni-kassel.de

Montesano, Francesco
National Research Council (CNR) - Institute of Sciences
of Food Production (ISPA), Italy
francesco.montesano@isp.cnr.it

Muchiri, Edward W.
Egerton   University,   Department   of   Civil   &
Environmental Engineering, Kenya
edmuchiri@yahoo.com

Nicolaos, M.
Faculty of Agricultural Technology, Technological Educational Institute of Epirus (TEI of Epirus), Greece

Ouamane, Tarek A.
Technical and Scientific Research Center on Arid Regions (CRSTRA), Biskra, Algeria
mailtoaltarek07@yahoo.fr

Pantelis, B.
Faculty of Agricultural Technology, Technological Educational Institute of Epirus (TEI of Epirus), Greece

Parente, Angelo
National Research Council (CNR) - Institute of Sciences of Food Production (ISPA), Italy
angelo.parente@ispa.cnr.it

Perner, Henrike
Leibniz-Institut of Vegetable and Ornamental crops, Großbeeren, Germany
perner@igzev.de

Henrike Perner studied at the University of Hohenheim agricultural biology and gained her PhD in agriculture in 2006 at the University of Berlin, Germany. Her research interest lays in plant nutrition and beneficial microorganisms. She worked in several international projects in the fields of recultivation of contaminated sites, food quality and organic farming. Currently she is working at the Institute of Vegetable and Ornamental Crops

in Großbeeren on the transfer of a water competence model into the agricultural sector, a project on vocational and educational training in Europe.

Psochios, Yannis
Agro-Know Technologies, Athens, Greece
psochios@agroknow.gr

Yannis is an Agro-Know partner. He has a lifetime involvement with the food and beverages sector where he has worked for years, also holding a specialization degree on their management and logistics. He has long working experience supporting SMEs that operate in the agricultural, food, beverages and tourism sectors in rural areas like Epirus and Crete, Greece.
Yannis has a particular interest in the way that ICT and Web technologies can be applied in real-life situations and this lead him to get involved in Agro-Know Technologies in 2008. For a long period of time he has been serving as the Managing Director of Agro-Know getting involved in to the administration & coordination of its participation in projects like WACOM, eCOTOOL, POLITICS and CerOrganic. He is currently coordinating the Outreach, Marketing & PR team in order to put in place activities that will bring Agro-Know in better contact with its targeted user audiences. He particularly enjoys travelling in various rural areas and islands in order to work close with local stakeholders and communities.

Rea, Elvira
Consiglio per la Ricerca e la sperimentazione in Agricoltura – Centro di Ricerca per lo studio delle relazioni tra Pianta e Suolo, Roma, Italy
elvira.rea@entecra.it

Rodriguez, Xenia
University of Duisburg-Essen, Campus Essen
Information Systems for Production and Operations
Management
xenia.rodriguez@icb.uni-due.de

Xenia Rodriguez is Project - and Finance Manager, working at the University of Duisburg-Essen, Campus Essen. Before managing the WACOM (Water Competences Model) Project and supporting the eCOTOOL (eCompetences Tools) Project she studied Business Economics (Bachelor of Science) at the University of Duisburg-Essen, Campus Essen (2011). Her research interests are E-Learning, Competence- and HR-Management with regard to cultural and international challenges as well as the Corporate Governance Debate (Master Studies at the University of Wuppertal since 2012).

Schwarz, Dietmar
Leibniz-Institut of Vegetable and Ornamental crops,
Großbeeren, Germany
schwarz@igzev.de

Dietmar Schwarz is a senior scientist at the Leibniz Institute for Vegetable and Ornamental Crops, Großbeeren, Germany since 1990. Before, he worked in the field of plant protection and as a horticultural farmer. Within his scientific career he focused on plant root research, protected cultivation in greenhouses, and vegetable quality. Beside his research he teaches plant nutrition, hydroponics, and organic horticulture and acts as one of the editors-in-chief for Scientia Horticulturae. He (co)authored more than 40 scientific publications, contributed to 6 book chapters, wrote many conference proceedings and publications for commercial horticulture. Within his international projects he worked for longer periods in labs in the Netherlands, USA, Israel und Mexico. The last years he gained experience in grafting solanaceae and published his work about the advantages and mechanisms of grafted tomato alleviating abiotic stresses. The focus of the

work is on temperature but other factors, such as radiation, nutrition (N, K, Al, pH) and salinity are also included.

Stracke, Christian M.
University of Duisburg-Essen, Campus Essen
Information Systems for Production and Operations Management
christian.stracke@icb.uni-due.de

Christian M. Stracke is Adjunct Professor at the Korea National Open University (KNOU) in Seoul and Coordinator of European research projects at the University of Duisburg-Essen. As an internationally recognized researcher his main working fields are competence modelling, technology enhanced learning, quality management and evaluation as well as standardisation in these fields. He is elected officer at the international (ISO-Convener for SC36) and European (Chair of CEN/TC 353) Standardisation Committees for TEL. He has published numerous articles and book chapters and is an invited speaker and PC member at many international conferences. As senior consultant and project manager he has gained extensive experience in leading projects and in consulting and supporting enterprises and public organisations to develop long-term policies and to implement and improve technology enhanced learning. Christian M. Stracke is the Co-Author of several standards for technology-enhanced learning and for competencies in HR with long-term experiences in their adaptation, application, and implementation within enterprises as well as research projects.

Švecová, Eva
Dipartimento di scienze e tecnologie per l'Agricoltura, le Foreste, la Natura e l'Energia, Università della Tuscia, Viterbo, Italy
eva.sve8@gmail.com

Thanopoulos, Charalampos
Agro-Know Technologies, Athens, Greece
cthanopoulos@agroknow.gr

Charalampos Thanopoulos has a diploma in Crop Science, Specialization in Vegetable Crop Production, a M.Sc. in Modern Systems of Crop Science, Plant protection and Landscape architecture (Specialization in Plant Physiology of Vegetables) and a Ph.D. in pre- / post-harvest physiology of vegetables, all from the Laboratory of Vegetable Production of the Agricultural University of Athens, Greece.

Babis (his short name) is serving as a trainer of adults / farmers, certified by the Accreditation Centre for Continuing Vocational Training (EKEPIS), in topics of vegetables cultivation (conventional & organic cultivation and open field & greenhouse production) for the Institute of Agricultural Sciences (Legacy of Ifigenia Sygros, Ministry of Rural Development and Food).

He is involved as an associate researcher for Agro-Know Technologies, Greek Research & Technology Network and Laboratory of Vegetable Production of the Agricultural University of Athens in several EU projects (agINFRA, AGRICOM, Organic.Lingua, WACOM, CerOrganic, eCOTOOL I.S.L.E. Erasmus Network,Rural Inclusion, LaProf, Organic.Edunet, bio@gro, Organic.Balkanet, V-3DAS).

Babis has strong experience in the development of learning objects (training material) in agricultural topics (bio@gro), implementation of metadata schemas for the description of learning resources (Organic.Edunet), exploitation of digital repositories (Organic.Edunet, Organic.Lingua, agINFRA), creation of educational and training scenarios in agricultural topics (Organic.Edunet, CerOrganic, Organic.Balkanet, V-3DAS), identification of competences for professional training (WACOM, eCOTOOL, CerOrganic, AGRICOM), design and development of Application Profiles for the online representation of job profiles and training opportunities (CerOrganic, WACOM, eCOTOOL, AGRICOM).

Tsirogiannis, Y. L.
Faculty of Agricultural Technology, Technological
Educational Institute of Epirus (TEI of Epirus), Greece

Venezia, Accursio
Consiglio per la Ricerca e la Sperimentazione in
Agricoltura - Centro di Ricerca per l'Orticoltura / Italian
Agriculture Research Council - Vegetable Crops
Research Centre, Pontecagnano (SA) Italy
accursio.venezia@entecra.it

AGRICOM Project Consortium Information

Partner and Acronym	Logo	Contact
University of Duisburg-Essen (Coordinator) *UDE*	UNIVERSITÄT DUISBURG ESSEN *Open-Minded*	**Christian M. Stracke** christian.stracke@uni-due.de
Leibniz-Institute of Vegetable and Ornamental Crops *IGZ*	IGZ	**Dr. Dietmar Schwarz** schwarz@igzev.de
International Foundation for Sustainable Agriculture Training *IFSAT*	IFSAT	**Bas Timmers** bastimmers@online.nl
Van der Meer & van Tilburg BV *VDM*	VAN DER MEER & VAN TILBURG	**Arjan De Bruin** deBruin@innovation.nl
Agro-Know Technologies *Agro-Know*	AGRO-KNOW TECHNOLOGIES	**Charalampos Thanopoulos** cthanopoulos@agroknow.gr
Agricultural University of Athens *AUA*	Agricultural University of Athens	**Dr. Dimitrios Savvas** dsavvas@aua.gr
University of Tuscia *UNITUS*	UNIVERSITÀ DEGLI STUDI DELLA Tuscia	**Giuseppe Colla** giucolla@unitus.it

About the
European project
AGRICOM:

Education and Culture DG

Lifelong Learning Programme

Main goal of the project AGRICOM is the transfer and population of the WACOM Water Competence Model (WCM) from the Water Sector to the Agricultural Sector.

Main objectives of AGRICOM are:

1. Identifying and analysing targeted needs and competences that are required by the labour market for specific use cases and jobs from several fields of the agricultural sector,
2. Transferring and adapting the generic WACOM competence model towards the AGRICOM competence model (ACM) allowing more agricultural use cases,
3. Pilot testing the ACM to the jobs specialisations related to agricultural uses of water resources (irrigation, hydroponics, etc.),
4. Establishing the AGRICOM online community for communication, moderated discussions and exchange of project results, experiences and expertise on competence modelling for vocational education and training in the agricultural sector,
5. Establishing the AGRICOM web portal and populating it with more agricultural use cases,
6. Enhancing the transparency of the job profiles in the agricultural sector,
7. Supporting the conjunction of vocational education and training with the labour market and jobs specialisations.

Main long-term objective and addressed impact is the introduction of competence modelling to the agricultural sector to develop a first Agriculture Competence Model and to strengthen the transparency and comparability of VET opportunities through the transfer of WCM and the adaptation of ECVET and EQF.

More information about AGRICOM online:
http://www.agriculture-competences.eu

AGRICOM Contact:

Coordinator: Christian M. Stracke
Organization: University of Duisburg-Essen
Address: Universitaetsstr. 9
                 45141 Essen, GERMANY
Telephone: +49 (0)201-183-4410
E-Mail: christian.stracke@uni-due.de

Lifelong Learning Programme

This project has been funded with support from the European Commission. This communication reflects the views only of the author, and the Commission cannot be held responsible for any use which may be made of the information contained therein.

Due to the shortage of the precious resource water and the consequences of the climate change, numerous European policies have been developed and adopted for the protection and sustainable utilisation of water creating a huge demand in particular in the vocational training. Economic factors like privatisation and increasing cost pressure in water management are sharpening these educational needs leading to the demand for specific Vocational Education and Training (VET) opportunities and products as short and tailor-made as possible.

The AGRICOM project supports close links to working life in order to make VET more responsive to the labour market's needs in the agricultural sector. AGRICOM facilitates and improves the identification and anticipation of skills and competences' needs and their integration in VET provision and implies also promoting integration of learning with working. In particular AGRICOM supports the implementation of the "New Skills for New Jobs" strategy by taking into account the challenges such as environmental and demographic changes and the related growing job needs also in the agricultural sector.

The presented articles are the result of the Open Call for Papers issued by the AGRICOM Consortium in order to raise the awareness of the stakeholders in the agricultural sector concerning competence modelling, with a special focus on hydroponics and irrigation.

The scientific articles published in this book are the selected papers of applicants from over six countries received upon the Open Call for Papers issued by the AGRICOM Conference 2013: They were reviewed by the scientific Programme Committee of AGRICOM 2013 in double-blind peer reviews and selected according the review results. In addition all authors of the selected articles could present and discuss their papers at the AGRICOM conference in a speech.

To summarise, this book contributes to the current developments and debate on competence modelling in the agricultural and horticultural sectors by presenting latest techniques, offering different views and solutions on competence modelling and by providing suggestions for future improvements European vocational education and training in the agricultural and horticultural sectors.

ISBN 978-3-8325-3540-7

Logos Verlag Berlin   **ISBN 978-3-8325-3540-7**

Abstract

In this work the importance of individualization in binaural technique is investigated. The results extend the present knowledge on the efficient measurement of individual head-related transfer functions (HRTFs) and highlight the importance of individual equalization filters in binaural reproduction, using both loudspeakers and headphones. Moreover, an integrated framework for the calculation of such equalization filters is presented.

An innovative measurement setup was developed to allow the fast acquisition of individual HRTFs. The hardware was designed to be compatible with the range extrapolation technique, which makes the description of the HRTF's distance-dependence possible. Major speedup was obtained by optimizing the multiple exponential sweep method. An individual HRTF dataset with approximately 4000 directions can be measured in less than 6 minutes with this new setup.

Crosstalk cancellation (CTC) filters are required when playing back binaural signals via loudspeakers. To allow listeners to freely move their heads, switching between multiple loudspeakers is required and the CTC filters must be constantly updated according to the tracked head position. Filter calculations are carried out in frequency-domain for speed reasons. To impose causality constraints to the regularized frequency-domain calculations, a CTC filter calculation framework was proposed, which incorporates a new approach for the multi-channel minimum-phase regularization. This framework also addresses the switching between active loudspeakers through the use of a weighted filter calculation. A sound localization test showed that individualized CTC systems provided performance similar to that from binaural listening while nonindividualized CTC systems provided significantly lower localization performance.

To deliver an authentic auditory impression without additional spectral coloration, binaural reproduction via headphones must be adequately equalized. Such equalization filters are obtained by inverting the headphone transfer function, which varies among listeners and individual fitting. To cope with these variations, a robust individual headphone equalization method was proposed. Perceptual tests showed that, in all but one of the tested situations, no audible differences between the original sound source and its binaural auditory display could be perceived.

Contents

List of Figures

List of Tables

List of Acronyms

3-AFC	three-alternative forced choice
AE	auditory event
ANOVA	analysis of variance
AP	all-pass
ATF	anatomical transfer function
BRvIEH	binaural reproduction via individually equalized headphones
CTC	crosstalk cancellation
CS	channel separation
$\widehat{CS}$	natural CS
DFT	discrete Fourier transform
DSP	digital signal processor
DTF	directional transfer function
DUT	device under test
EC	ear canal
ED	eardrums
FEC	free-air equivalent coupling
FIR	finite impulse response
FF	free-field
FFT	fast Fourier transform
HP	headphones
HpTF	headphone transfer function
HRIR	head-related impulse response
HRTF	head-related transfer function
IIR	infinite impulse response
ILD	interaural level difference
IR	impulse response
IT	interleaving

ITA	Institute of Technical Acoustics (RWTH Aachen University)
ITD	interaural time difference
LE	lateral error
LMS	least mean square
LSM	least-square minimization
LTI	linear time-invariant
MESM	multiple exponential sweep method
MIMO	multiple input, multiple output
MLS	maximum length sequence
MP	minimum-phase
OSD	optimal source distribution
OV	overlapping
PCA	principal component analysis
PDR	pressure division ratio
PE	polar error
PSD	power spectral density
QE	quadrant error
RIR	room impulse response
RM ANOVA	repeated-measures ANOVA
RMS	root-mean-square
SE	sound event
SISO	single input, single output
SH	spherical harmonic
SM	sequential method
SNR	signal-to-noise ratio
SPL	sound pressure level
TF	transfer function
VBAP	vector base amplitude panning
VR	virtual reality

List of Symbols

Mathematical Notation

$a(t)$	lowercase: signal in time-domain
$A(f)$	uppercase: signal in frequency-domain
$\boldsymbol{a}$	bold lowercase: vector in frequency-domain (unless stated otherwise)
$\boldsymbol{A}$	bold uppercase: matrix in frequency-domain (unless stated otherwise)

Mathematical Operators

$*$	convolution
$\circ$	element wise multiplication
$\lvert \cdot \rvert$	modulus of a complex number
$\lVert \cdot \rVert_2$	Euclidean-norm of a vector or matrix
$\lVert \cdot \rVert_{\boldsymbol{W}}$	weighted-norm of a vector
$(\cdot)^*$	conjugate of a complex number or adjoint operator (Hermitian transpose) of a vector or matrix
$(\cdot)^T$	transpose of a vector or matrix
$\mathrm{adj}(\cdot)$	adjugate operator of a matrix
$\det(\cdot)$	determinant of a matrix
$\mathrm{diag}(\cdot)$	diagonal matrix, whose diagonal entries are given by an input vector
$(\cdot)^+$	Moore-Penrose pseudo-inverse
mod	modulo operation
$\lceil \cdot \rceil$	ceiling operator (round up to next integer)
$(\cdot)^+$	minimum causal stable parts of a function

$(\cdot)^-$	minimum anti-causal stable parts of a function
$[\cdot]_+$	causal part of a function

Mathematical Symbols

$x(t)$	continuous time signal
T_s	sampling period
f_s	sampling frequency
$x(n)$	discrete input time vector
$y(n)$	discrete output time vector
$w(n)$	discrete time window vector
$h(n)$	impulse response of a system
$X(f)$	input spectrum
$Y(f)$	output spectrum
$X(z)$	discrete input spectrum
$H(z)$	transfer function of a system
$D(z)$	desired frequency response
$Q(z)$	equalization filter
T	linear transformation
$\delta(n)$	Kronecker delta function
$\boldsymbol{y}$	measurement vector
$\boldsymbol{A}$	transformation matrix
$\boldsymbol{x}$	vector of least square solutions
$\boldsymbol{I}$	identity matrix
$\mu(z)$	regularization parameter
$s(n)$	discrete time sine sweep
$S(z)$	sine sweep discrete spectrum
$\phi_\mathrm{sw}(n)$	sweep's phase increment
ϕ_0	sweep's starting phase.
$A_\mathrm{reg}(z)$	regularization filter
$H_\mathrm{L}(z)$	left channel HRTF

$H_{\mathrm{R}}(z)$	right channel HRTF
τ_{ikj}	travel time between the ith loudspeaker to the kth microphone at the jth measurement position
$\delta_{\lceil i/8 \rceil}$	latency of an eight channel sound card
x_i	position of the ith loudspeaker
P_{kj}	position of the kth microphone at the jth measurement position
r_k	distance between the kth microphone and the center of the bar holding the microphones
ϕ_j	angle of the turntable at the jth measurement position
d	distance between the bar holding the microphones and the stand connected to the turntable
ρ	angle between the bar holding the microphones and the stand connected to the turntable
γ	torsion of the bar hold the microphones
$\mathbf{x}$	position vector
T_{SM}	total measurement time for SM
T_{MESM}	total measurement time for MESM
τ_{sw}	length of excitation sweep
τ_{st}	length of stop margin
τ_{w}	waiting time
τ_{IR}	length of room impulse response
τ_{DUT}	length of desired impulse response
τ_{sp}	length of safety region
k	harmonic order
a_k	minimum difference between harmonic and fundamental spectrum
r_{sw}	sweep rate
Δt_k	distance between fundamental and k^{th} harmonic IR
$H_{\mathrm{free\text{-}field}}$	free-field equalized HRTF
H_{ref}	reference transfer function
H_{diff}	diffuse field HRTF
t_{win}^i	start time of a time window

$\boldsymbol{\theta}$	direction vector
p	pressure value
m	degree of SH
n	order of SH
l	linear SH-index ($l = n^2 + n + m + 1$)
O	truncation order
c_{nm}	complex spherical expansion coefficients
ξ_{nm}	spherical expansion coefficient
h_n	spherical Hankel function of order n
P_n^m	Legendre functions of order n and degree m
$Y_n^m(\boldsymbol{\theta})$	spherical harmonic of order n and degree m for direction $\boldsymbol{\theta}$
k	wavenumber ($k = 2\pi f/c$)
f	frequency
c	speed of sound
r	radial distance
ϕ	azimuth angle
θ	elevation angle
α	lateral angle
β	polar angle
$\boldsymbol{w}$	weight vector
$\boldsymbol{p}$	pressure vector
$\boldsymbol{Y}$	spherical harmonic basis functions matrix
$\boldsymbol{c}$	spherical expansion vector
$\boldsymbol{\xi}$	spherical potential vector
S	spherical surface
$H_{\mathrm{FF}}^{\mathrm{ED}}(z)$	transfer function from loudspeaker to eardrums
$H_{\mathrm{FF}}^{\mathrm{EC}}(z)$	transfer function from loudspeaker to the microphones at the entrance of the ear canal
$H_{\mathrm{HP}}^{\mathrm{ED}}(z)$	transfer function from the headphone to the eardrums
$H_{\mathrm{HP}}^{\mathrm{EC}}(z)$	transfer function from the headphone the microphones at the entrance of the ear canal.
L	number of loudspeakers

M	length of CTC filters
N	length of HRIR
$\boldsymbol{H}$	transfer matrix
$\boldsymbol{C}$	CTC matrix
$\boldsymbol{b}$	vector of binaural signals
$\boldsymbol{v}$	vector of transaural signals
$\boldsymbol{e}$	vector of ear signals
$\boldsymbol{d}$	vector of desired signals
Δ	CTC filter delay
$R(z)$	band-pass filter

1

Introduction

Spatial audio systems can, among other applications, be found in home entertainment, cinema and virtual reality systems where they are used to provide the listener with an increased sense of realism and immersion. RUMSEY (2012) lists some of the spatial attributes required for sound reproduction: naturalness, source localization, distance and depth perception, envelopment and spaciousness, and apparent source width.

There are several different approaches to the creation of spatial auditory impression. Amplitude panning techniques, such as the standard stereophony or its three-dimensional extension, the "vector base amplitude panning" (PULKKI, 1997) or "ambisonics" (as originally formulated by GERZON, 1973) are based on the psychoacoustic effects of summing localization (BLAUERT, 1997). These systems have the drawback that all produced phantom sources are perceived at the distance of the loudspeakers. Wave-field reconstruction methods such as "wave field synthesis" (DE VRIES, 1988) and "higher-order ambisonics" (DANIEL, 2003) focus on completely reconstructing a desired sound field inside the reproduction space. These systems require a large number of sources to work properly and even though they are theoretically capable of rendering focused sources, their realization is severely limited in practical applications.

LENTZ (2007) and SCHRÖDER et al. (2010) describe the virtual reality system at RWTH Aachen University. This system is composed of five video projection walls that severely limit the positioning of loudspeaker to be used with spatial audio generation. To ensure a spatial audio reproduction, thus increasing the immersion in the virtual environment, this system was designed to use yet another type of spatial audio reproduction method, the binaural technology.

The binaural technology can provide the listener with full three-dimensional impression, i.e. lateral position, height and distance impression, with a reduced number of transducers (MØLLER, 1992). Binaural technology is based on the fact that all spatial sound information perceived by humans is extracted solely from the two pressure signals captured by the listener's ears. These so-called binaural signals can be either directly recorded or they can be synthesized.

Binaural recordings are made using an artificial head or in-ear microphones (PAUL, 2009). They have, however, the disadvantage that listener's head movements cannot be directly compensated for. LI and DURAISWAMI (2006) proposed a method to binaurally "reproduce 3D auditory scenes captured by spherical microphone arrays over headphones". Nevertheless, virtual reality applications commonly rely on binaural synthesis, which will be the focus of this work.

In the binaural synthesis, each sound source must be filtered with a head-related transfer function (HRTF) that describes the direction-dependent influence of pinna, head, and torso on the incident sound field. HRTF are, therefore, listener-dependent (WIGHTMAN and KISTLER, 1989; WENZEL et al., 1993; MØLLER et al., 1995a), and, when using binaural technology to create an authentic spatial sound scene, individual HRTFs should be used. As a binaural synthesis is based on acoustic simulations, its performance also depends on the quality of the used acoustic model.

The first part of this work will focus on the characterization of the human listener, i.e. on the acquisition of individual HRTFs. One possibility to acquire HRTFs is to use numerical acoustic simulations, like the boundary element method (KATZ, 2001) or the finite difference time-domain (MOKHTARI et al., 2007), based on mesh grids obtained from e.g. magnetic resonance imaging (MOKHTARI et al., 2007; GUILLON et al., 2012), laser scanning (RUI et al., 2012) or photogrammetry (FELS, 2008; DELLEPIANE et al., 2008). However, as the outer ear contains hidden structures that contribute to the HRTF at high frequencies, the quality of such method is still not ideal. HRTFs can also be obtained through acoustic measurements. This can be done either in a direct (BRONKHORST, 1995; MØLLER et al., 1995a; ALGAZI et al., 2001; MAJDAK et al., 2007) or in a reciprocal manner (ZOTKIN et al., 2006). No matter which method is used, it is desirable that the HRTF measurement is completed in the shortest time possible, thus providing more comfort for the subject being measured and reducing measurement variability.

The presentation of near-to-head sources can greatly improve the naturalness of virtual reality systems and the synthesis of near-to-head sources requires near-field HRTFs (BRUNGART and RABINOWITZ, 1999). Near-field HRTFs have to be either measured using a complex setup that is composed of transducers placed at increasing distances, or they can be calculated from a measurement at a single distance using the range extrapolation technique (DURAISWAMI et al., 2004).

The second part of this thesis will deal with the reproduction of binaural signals. The main requirement for a binaural reproduction is that each of the listener's ears receives a distinct signal. Such a perfect channel separation can easily be achieved by reproducing binaural signals via headphones. However, headphones introduce spectral coloration and may change the acoustic impedance seen from the ear canal, influencing the naturalness of the binaural reproduction. The variation in impedance can be only controlled by choosing the correct headphone types (MØLLER, 1992). The coloration aspect, which is highly individual and dependent on the headphone fitting (MØLLER et al., 1995b), can be compensated for by using an individual equalization filter (PRALONG and CARLILE, 1996; MØLLER et al., 1995b).

If a pair of loudspeakers is used to directly reproduce a binaural signal, the sound radiated from each loudspeaker will arrive at the listener's ears, thus mixing the binaural cues contained in the binaural signal. To reestablish these cues, crosstalk must be eliminated (BAUER, 1961; ATAL et al., 1966). This is achieved by using a crosstalk cancellation (CTC) filter network (BAUCK and COOPER, 1993; KÖRING and SCHMITZ, 1993; KIRKEBY and NELSON, 1999).

The design of CTC filters depends on the disposition of the loudspeakers in relation to the listener. If the listener is located in a limited region in the reproduction space, then the aim of the CTC filter design is to provide a wide sweet spot. This can be optimally achieved with a frequency-dependent loudspeaker distribution (TAKEUCHI and NELSON, 2007). On the other hand, if the listener should be allowed to freely move inside the reproduction space, then the sweet spot has to be made dynamic by constantly updating the CTC filters based on the listener's current tracked position (GARDNER, 1997; LENTZ, 2007). To avoid filter instability, the dynamic CTC system should switch between ideal loudspeaker configurations, dependent on the listener's position and direction (LENTZ, 2006).

As CTC filters are designed based on the transfer function between loudspeakers and listener's ears, i.e. HRTFs, individualized CTC filters provide a higher channel separation than its generic counterpart (AKEROYD et al., 2007). It is, however, not yet known how a reduced channel separation influences the localization performance of a nonindividualized CTC system.

1.1 Objectives

The global aim of this work is to improve the quality of the binaural technology used in the virtual reality system at RWTH Aachen University by means of individualization. Therefore only the single listener situation will be considered.

More specifically, the following questions will be discussed in this thesis:

- How to acquire individual HRTFs in a fast manner with full 3D information?

- How important is the individual equalization for a binaural reproduction?

- How can a binaural reproduction be equalized adequately?

1.2 Organization

First, some fundamental aspects of digital signal processing, acoustic measurement and spatial hearing, required for the better comprehension of this work, are presented in chapter 2. The characterization of human listeners for binaural synthesis, i.e. the measurement of individual HRTFs, is discussed in chapter 3, where the design of a measurement system for individual HRTFs is presented, followed by the optimization of the used excitation signals and a discussion on the conducted post-processing. The chapter concludes presenting measurement results obtained using the described system.

The focus is then put on the reproduction of binaural signals. In chapter 4, a reproduction via headphones is examined, testing the adequacy of the of two headphone types for binaural reproduction and presenting a framework for the calculation of individual equalization filters. Binaural reproduction via loudspeakers is then the focus of chapter 5, where a framework for the calculation of individual causal crosstalk cancellation filters in the frequency-domain is presented, which also incorporates the switching between active loudspeakers in 360° scenarios.

The importance of individually equalized binaural reproduction is investigated in chapter 6 by means of two perceptual tests. Section 6.1 evaluates the naturalness of individually equalized headphone binaural presentation while section 6.2 investigates the localization performance of individualized and nonindividualized crosstalk cancellation systems. At the end, chapter 7 summarizes the results and contributions presented in this thesis and possible future research is discussed.

2

Fundamental Concepts

In this chapter some fundamental aspects required for the remainder of this thesis will be briefly summarized. First, a review of the topic of *digital signal processing*, is presented. Digital signal processing is essential when dealing with acoustic measurements, such as the head-related transfer function measurement discussed in chapter 3, and filter designs, such as the headphone equalization and the crosstalk cancellation filters described in chapter 4 and chapter 5.

Both, acoustic measurements and filter design, and the acoustic holography discussed in chapter 3, can be considered as inverse problems. Therefore, some important aspects of how to solve inverse problems are discussed. Acoustic measurements are further analyzed with regard to the used excitation signal and deconvolution methods.

To conclude, a short review of the human directional hearing is presented as the perceptual evaluation presented in chapter 6 is based on it.

2.1 Digital Signal Processing

An acoustic signal can be represented as a real continuous time function $x(t)$. The signal must be discretized so that is can be *digitally* processed. The continuous signal is sampled at a regular interval T_s (reciprocal to the sampling frequency f_s). The result is a sequence of values x_n, where n is the time index. The notation $x(n)$ is used for the vector containing the values of x_n (OPPENHEIM and SCHAFER, 1989).

TOHYAMA and KOIKE (1998) state that a discrete system is "a system that transforms an input sequence $x(n)$ in an output sequence $y(n)$." Such a system can be described by its impulse response (IR) sequence $h(n)$. In this thesis all systems are considered *linear and time invariant* (LTI, OPPENHEIM and SCHAFER, 1989, p. 22). The output sequence of an LTI system is derived from the input and the system's IR by performing the convolution operation

$$y(n) = x(n) * h(n) = \sum_{k=-\infty}^{\infty} x(n)\,h(n-k). \tag{2.1}$$

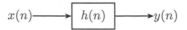

Figure 2.1: A system described by its IR $h(n)$ with input sequence $x(n)$ and output sequence $y(n)$.

The IR of most physical systems is infinitely long and can be described by an *infinite impulse response* (IIR) filter. However, these systems are commonly approximated by a *finite impulse response* (FIR) filter, which allow a more efficient calculation of the convolution operation (OPPENHEIM and SCHAFER, 1989).

A finite IR is usually obtained by truncating an infinite IR. To avoid an abrupt end to the signal, the *windowing* operation can be applied, which performs an element-wise multiplication of the sequence with a window sequence whose elements are zero (or close to zero) at the regions of unwanted components and therefore

$$x_{\text{win}}(n) = x(n) \circ w(n). \tag{2.2}$$

OPPENHEIM and SCHAFER (1989) show that complex exponential functions are eigenfunctions of LTI systems. Thus, a representation of signals as these complex exponentials, i.e. sinusoids, is very beneficial as a faster calculation of the convolution operation is possible in this domain. A representation of a signal that uses complex exponentials as basis is said to be in the *frequency-domain* while the previous representation is said to be in the *time-domain*. In the frequency-domain, a system is described by its frequency response $H(f)$. The output sequence is obtained from

$$Y(f) = H(f) \cdot X(f). \tag{2.3}$$

$Y(f)$ and $X(f)$ are the complex frequency spectrum of $y(n)$ and $x(n)$ respectively.

Time and frequency representation of a signal are closely related. If the spectrum is, for instance, real, the time representation is symmetric around the origin and noncausal. Furthermore, under certain conditions, the spectrum's magnitude and phase can also be dependent on each other. Therefore, a minimum-phase LTI system—a stable and causal system whose inverse is also stable and causal—has an amplitude and phase that are related by the *Hilbert transform* (OPPENHEIM and SCHAFER, 1989, Ch. 11).

A transformation from time into frequency-domain is obtained using the *Fourier transform.* For finite length sequences, however, the *discrete Fourier transform* (DFT) is usually preferred as efficient algorithms exist for its calculation, the so-called *fast Fourier transform* (FFT). The output of the DFT is itself a sequence of samples, equally spaced in frequency, of a signal's Fourier transform (OPPENHEIM and SCHAFER, 1989). In the following, the term frequency-domain will refer to the output of a DFT and the DFT from $x(n)$ will be denoted by $X(z)$, the discrete representation of $X(f)$.

The DFT is based on the assumption that the signal is a periodic sequence. Thus, periodicity in both time and frequency-domain is inherent to the DFT and caution is required when conducting some operations with a DFT. It is especially important to note that the linear convolution operation is redefined as a circular convolution under the DFT. This can lead to the presence of time-aliasing if the length of both signals to be convolved is not large enough (OPPENHEIM and SCHAFER, 1989). If the result of the linear convolution presents noncausal components, Using the circular convolution these components will now appear at the end of the output sequence, an effect know as *wrap around.*

The shifting operation, which can be understood as a convolution with an impulse shifted in time, must also be redefined as a circular shift. This means that when shifting a sequence by N samples, the last N samples of the sequence are simply moved to the beginning of the sequence.

2.2 Inverse Problems

All signal processing algorithms presented in this thesis can be interpreted as *inverse problems.* Inverse problems occur when a desired signal cannot be measured directly. In this case, the acquired signals are a transformation of the desired signals and an inversion of these transformations is thus required (MAMMONE, 1999).

Measurements of acoustic systems play a major role throughout this thesis. As explained in section 2.3, measurements of acoustic systems are conducted using deterministic excitation signals rather than an ideal impulse. It is assumed that the excitation signal $x(n)$ is the output of a transformation T to the Kronecker delta function $\delta(n)$. Thus, the system's IR can be obtained from the system's output $y(n)$ by applying the inverse transformation T^{-1} to $y(n)$.

The equalization problem can also be considered as an inverse problem. In this case, the frequency response of the equalization filter $Q(z)$ is obtained from the difference between the desired overall response $D(z)$ and the system's equalized frequency response. The ideal equalization filter is obtained when the observed difference between the desired frequency response and the equalized frequency response is driven to zero, i.e.

$$|D(z) - Q(z)H(z)| = 0, \tag{2.4}$$

$$Q(z) = D(z)/H(z). \tag{2.5}$$

The headphone equalization filter calculation discussed in chapter 4 is a classic example of a *single input, single output* (SISO) equalization problem while the crosstalk filter design discussed in chapter 5 is an example of a *multiple input, multiple output* (MIMO) equalization problem.

Finally, the acoustic holography used for interpolation and range extrapolation in chapter 3 is also an inverse problem as the desired potential expansion coefficients cannot be directly measured and must be estimated from the pressure measurement over a spherical surface by inverting the spherical harmonic transformation.

The inverse problems cannot always be directly inverted as in eq. (2.5). This is especially true for the MIMO cases, which can have an over- or underdetermined transformation matrix. These problems can, however, be considered a minimization problem of the form

$$\min_{\boldsymbol{x}} \|\boldsymbol{y} - \boldsymbol{A}\boldsymbol{x}\|_2^2, \tag{2.6}$$

where $\|\cdot\|^2$ is the Euclidean norm, $\boldsymbol{y}$ is a vector containing the measured values, $\boldsymbol{A}$ is the transformation matrix, and $\boldsymbol{x}$ is the solution of the inverse problem.[1] The solution to eq. (2.6) is given by

$$\boldsymbol{x} = \boldsymbol{A}^+ \boldsymbol{y} = \begin{cases} (\boldsymbol{A}^*\boldsymbol{A})^{-1}\,\boldsymbol{A}^*\boldsymbol{y}, & \text{if } \boldsymbol{A} \text{ is overdetermined} \\ \boldsymbol{A}^{-1}\boldsymbol{y}, & \text{if } \boldsymbol{A} \text{ is invertible} \\ \boldsymbol{A}^*\,(\boldsymbol{A}\boldsymbol{A}^*)^{-1}\,\boldsymbol{y}, & \text{if } \boldsymbol{A} \text{ is underdetermined} \end{cases} \tag{2.7}$$

where $\boldsymbol{A}^*$ represents the Hermitian transpose of matrix $\boldsymbol{A}$ and $\boldsymbol{A}^+$ its Moore-Penrose pseudo-inverse assuming that $\boldsymbol{A}^*\boldsymbol{A}$ and $\boldsymbol{A}\boldsymbol{A}^*$ are not singular.

---

[1] Even though the norm to be minimized does not necessarily have to be the Euclidean norm, this norm will be used throughout this work, as it has an analytical solution derived from the *least-squares minimization* (LSM) when $\boldsymbol{A}$ is overdetermined and from *constraint optimization* when $\boldsymbol{A}$ is underdetermined (NELSON and ELLIOTT, 1995, pp. 418–419).

It is often the case that the observed data contains additive noise and that the transformation matrix is not well-conditioned. This leads to an undesirable noise amplification effect that can severely affect the quality of the obtained solution (FAZI, 2010). Therefore, regularization can be introduced to mitigate the minimization problem and avoid over-fitting. The most commonly used regularization scheme is the Tikhonov regularization that adds a restriction on the total energy of the solution vector.

In this case, the transformation matrix A is overdetermined, the minimization eq. (2.6) is altered to

$$\min_{x} \left(\|y - Ax\|_2^2 + \mu \|x\|_2^2 \right), \tag{2.8}$$

where μ is the regularization parameter with real values in the range $0 \leq \mu \leq \infty$. μ acts as a trade-off factor between the residual error and energy of the solution. Moreover, for the equalization problems, μ also acts as an upper limit of the resulting filter's maximum gain (see appendix A).

The regularization has, however, its drawbacks. The use of regularization to solve equalization problems leads to ringing artifacts in the time-domain, as discussed by BOUCHARD et al. (2006) for the one channel case and by NORCROSS and BOUCHARD (2007) for the multi-channel case.

The solution of the new minimization problem eq. (2.8) is given by

$$x = \begin{cases} (A^*A + \mu I)^{-1} A^* y, & \text{if } A \text{ is overdetermined} \\ A^* (AA^* + \mu I)^{-1} y, & \text{if } A \text{ is underdetermined} \end{cases}, \tag{2.9}$$

where I is the identity matrix.[2] The solution of the underdetermined case is thoroughly explained in appendix B.

2.3 Acoustic Measurement

Acoustic measurements focus either on acoustic signals or on acoustic systems. The measurement of acoustic signals is common in fields such as noise control, e.g. to define how "loud" an acoustic source is, or to describe how "loud" it is inside a room, or even to define how long

---

[2]In case A is a square matrix, any of the two solutions can be used.

a person can be exposed to a given noise. Measurements of acoustic signals can also focus on the binaural technology. In this case, the recording is carried out using either in-ear microphones or an artificial head (PAUL, 2009). Binaural recordings, however, are not covered in this publication. Therefore, the term *acoustic measurement* will in the following always refer to the measurement of an acoustic system, in particular the measurement of the head-related transfer function (HRTF), as described in chapter 3.

Acoustic systems are assumed to the linear and time-invariant and can thus be described by its impulse response (IR) or by its frequency equivalent transfer function (TF).[3] If the spatial characteristic of an acoustic system is described, e.g. the directivity of a loudspeaker or a directional microphone, then a series of IRs measured at different directions will be required. The IR could be measured directly by feeding the system with an infinitely narrow impulse and recording the systems response. However, an impulse contains very little energy. When measuring under normal conditions, i.e. with background noise present, the obtained signal-to-noise ratio (SNR) will be low and many repetitions of the measurement will be necessary for an improved SNR (MÜLLER and MASSARANI, 2001).

To avoid repeated measurements, the *correlation technique* is commonly applied. This technique allows measurements to be conducted with any type of excitation signals, e.g. white noise, pink noise, or music. Nevertheless, some signals may be more suitable than others as their energy is better distributed over time and, therefore, yield a better SNR. Several different excitation signals have been studied regarding their performance in terms of SNR, crest factor, and measurement duration. From all these signals, sine sweeps and pseudo-random sequences are commonly preferred as such deterministic excitation signals provide high measurement repeatability (MÜLLER, 2008).

As discussed in section 2.2, the TF of an acoustic system is obtained by dividing the output spectrum of the system under test by the spectrum of the input signal. In time-domain, this division is equivalent to filtering the output signal with a *matched filter*, which is itself equivalent to calculating the signals' *cross-correlation* (MÜLLER, 2008). Therefore, a class of binary signals exhibiting unity auto-correlation was commonly

---

[3]According to MÜLLER (2008), "the IR can be transformed into the TF via Fourier transform (see section 2.1) and back again into the IR via the inverse Fourier transform, both are equivalent and carry the same information, which can be extracted and visualized in different ways." Thus, the terms IR and TF will be used in reciprocal form throughout this thesis.

employed in acoustic measurements. As listed by PELTONEN (2000), Golay codes (GOLAY, 1961), Legendre sequences (SCHROEDER, 1979), Barker codes (KUTTRUFF, 2000), and *maximum length sequences* (MLS) (BORISH and ANGELL, 1983) are examples of such sequences. The MLS technique became particularly popular as its auto-correlation can be calculated in an extremely efficient manner—in terms of computation time and memory usage—using the fast Hadamard transform. The main drawback of using pseudo-random sequences is, however, that this technique is very sensitive to time-variance and nonlinearity in the measurement chain (MÜLLER and MASSARANI, 2001).

As the calculation time of IRs became less critical due to the improvement of the calculation complexity and speed of state-of-the-art personal computers, sweep measurements became increasingly popular. Sweeps offer great advantages for systems that do not fully comply with the LTI assumption, e.g. weak time variances (slow changes of the system response over time) or nonlinear transfer characteristics (harmonic distortion) (MÜLLER and MASSARANI, 2001), as it can still achieve high SNR where pseudo-random sequences would succumb.

2.3.1 Exponential Sweep

Sweeps, also known in literature as chirp or swept-sine, are generally defined in time-domain as

$$s(n) = \sin(\phi_{\text{sw}}(n) + \phi_0) \tag{2.10}$$

where $\phi_{\text{sw}}(n)$ is the phase increment and ϕ_0 the starting phase (MÜLLER and MASSARANI, 2001). It is convenient to set the starting phase to $\phi_0 = 0$ as this results in a smooth start of the signal.

The two most commonly known types of sine sweeps are the linear and the exponential sweep. They differ in terms of how $\phi_{\text{sw}}(n)$ varies with time. They are aptly named as the former varies linearly in time while the later varies exponentially in time.

Exponential sweeps, when compared with linear sweeps, are known to have an advantage when it comes to non-linear systems, besides they contain higher energy at the lower frequency range, which is exactly the region where measured SNR tends to be worse. MÜLLER and MASSARANI (2001) show that the nonlinear behavior a system, when measured with an exponential sweep, can be observed as anti-causal IRs for different harmonic orders k,[4] each having a length $\tau_{\text{IR},k}$ and

---

[4]The harmonic order k starts at 2, as $k = 1$ is the desired fundamental IR itself.

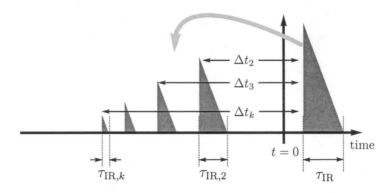

Figure 2.2: IR of a weakly nonlinear system obtained by exponential sweep measurement.

lagging Δt_k from the fundamental IR. The logarithmic magnitude of a weakly nonlinear system's IR measured with an exponential sweep is displayed schematically in fig. 2.2. Unless distortion measurements are being conducted, the fundamental IR (located to the right side in this example) is the result that we are actually interested in.

2.3.2 Regularized Deconvolution

When dividing the output spectrum by the input spectrum problems may arise for frequencies where the values of the exponential sweep $S(z)$ become very small. This would also be the case if the sweep covers only a limited bandwidth. The division of the output spectra by these small values can amplify the additive noise commonly present at the obtained system's output. This leads to undesirable artifacts in the IR.

In order to solve this problem, FARINA (2007) introduced the Tikhonov regularization for sweep measurements. The regularized inverse sweep is thus obtained by applying eq. (2.9) to the excitation sweep. Multiplying the matched filter with the system's output results in the regularized transfer function

$$H_{\text{reg}}(z) = \frac{Y(z)}{S(z)} \frac{1}{1 + \mu(z)/|S(z)|^2} = \frac{Y(z)}{S(z)} \cdot A_{\text{reg}}(z), \qquad (2.11)$$

where $A_{\text{reg}}(z)$ describes the influence of the regularization as a filter. Note that the regularization parameter $\mu(z)$ is now frequency-dependent.

In this case, if the exponential sweep is band-limited in $[f_1, f_2]$, it is reasonable to keep $\mu(z) = 0$ inside this range and to set $\mu(z) \gg \max |S(z)|^2$ elsewhere. By doing so, $A_{\text{reg}}(z)$ will suppress the measurement noise outside the desired frequency range.

Note, however, that $A_{\text{reg}}(z)$ is a zero-phase band-pass filter, which means that its temporal counterpart $a_{\text{reg}}(n)$ is symmetric regarding the time axis. As described by BOUCHARD et al. (2006), this leads to the noncausal behavior observed in IRs obtained using the regularized deconvolution.

Minimum-phase Regularization

To solve the problem of noncausality, $A_{\text{reg}}(z)$ can be factorized into a minimum-phase (MP) regularization filter and a remaining noncausal all-pass (AP) filter (TOHYAMA and KOIKE, 1998; BOUCHARD et al., 2006):

$$A_{\text{reg}}(z) = A_{\text{reg,MP}}(z) \cdot A_{\text{reg,AP}}(z). \tag{2.12}$$

If $A_{\text{reg,MP}}$ is used in the deconvolution process, the resulting fundamental IRs obtained when measuring a physical (and therefore causal) system will also be causal. This is especially important for the multiple exponential sweep method presented in chapter 3. For this technique noncausal IRs are problematic as the pre-ringing from one IR can overlap with a previous IR, which leads to artifacts when the IRs are cropped out of the raw IR. The phase error introduced in the pass-band by the minimum-phase filter can be later compensated by filtering the resulting IR with the all-pass component $A_{\text{reg,AP}}(z)$, yielding back noncausal IRs.

2.4 Directional Hearing

The spatial impression perceived by human beings is based on cues imprinted on the signal arriving at the listener's ear, the so-called *binaural signal*. These cues are caused by an alteration on the incoming wave front due to reflection, diffraction, shadowing, resonance, and dispersion at the listener's body (mainly torso, head, and pinna), which means that these cues are highly individual (WIGHTMAN and KISTLER, 1989; WENZEL et al., 1993; MØLLER et al., 1995a; FELS, 2008).

MØLLER et al. (1996) analyzed the localization performance of subjects with their individual and nonindividual binaural recordings. They concluded that "when compared to real life, the localization performance was preserved with individual recordings. Nonindividual recordings resulted in an increased number of errors for the sound sources in the median plane".

Similar results have also been reported for synthesized binaural signals. WENZEL et al. (1993) analyzed the localization performance of 16 subjects presented with the HRTF of another representative subject. They concluded that "while the interaural cues to horizontal location are robust, the spectral cues considered important for resolving location along a particular cone-of-confusion are distorted by a synthesis process that uses nonindividualized HRTFs."

MIDDLEBROOKS (1999b) compared the localization performance obtained with a loudspeaker (the natural condition) and with individual and nonindividual synthesized auditory displays. He verified that "performance in the own-ear virtual (individually synthesized) condition was nearly as accurate as that in the free-field (natural listening) condition." Furthermore, he reported that "all error measures of RMS errors tended to increase with increases in the spectral difference between the listener's (directional transfer function) DTFs and the DTFs used in the localization trials".

MIDDLEBROOKS (1999b) went further and scaled the nonindividual DTFs to make it more similar to the listeners own DTF. This alteration resulted in an improved localization performance, indicating the possibility of creating "a realistic virtual synthesis of auditory space for the large number of listeners for whom it would not be practical to make individual acoustical measurements of DTFs."

The results listed above indicate the importance of individual—or at least individualized—binaural synthesis. It is important to mention that all these results were obtained with inexperienced listeners without a training period. HOFMAN et al. (1998) demonstrated "the existence of ongoing spatial calibration in the adult human auditory system", i.e., they showed that human listeners are able to learn how to hear with a different ear. Another very important conclusion obtained by them was that "learning the new spectral cues did not interfere with the neural representation of the original cues, as subjects could localize sounds with both normal and modified pinnae" after a sufficiently long training period. Specifically for sound localization tests, MAJDAK et al. (2010) showed the importance of training as subjects "learn to better localize sounds in terms of precision, bias, and quadrant error".

Specifically regarding virtual reality applications, BEGAULT et al. (2000) argued that besides individual HRTFs, addition of head-tracking significantly decreased quadrant confusion and the addition of reverberation significantly improved the impression of externalization.

2.4.1 Head-Related Transfer Functions

The alterations caused at the wave front arriving from any given direction can be described by a pair of filters, frequently called *head-related transfer functions* (HRTF, BLAUERT, 1997), but also known as *anatomical transfer function* (ATF, HARTMANN, 1999). There are two transfer functions for each direction of sound incidence ($H_L(z)$ for the left and $H_R(z)$ for the right ear), which are combined into one HRTF.

An example of a measured HRIR is shown in fig. 2.3. This HRIR was measured with the sound source positioned at the right side of the listener. This information can be easily extracted from this plot, as the right signal is louder than the left signal, as this is attenuated by the head shadowing and it also arrives earlier than the left signal, as the acoustic path between the source and the left ear is longer than to the right ear.

The amplitude of the equivalent HRTF is shown in fig. 2.4. Once again, it is easy to verify that the source is at the listener's right side, as the right signal is louder throughout the whole human listening range. It is also possible to verify that this level difference is more pronounced in the higher frequency range.

The resonance behavior observed at fig. 2.4 can be associated to anthropometric characteristics of the human listeners. At low frequencies very low variance is observed, as for these frequencies the head and torso are small when compared to the wavelength. The shoulder reflection will create a comb filter effect, with the lowest resonance occurring in the region of 1.5 kHz and repeating every 3 kHz. The influence of the pinnae in the HRTFs is significant only to frequencies above approximately 2 kHz. A constructive resonance in the pinna results in a global maximum at approximately 5 kHz. A sharp minimum in frequencies around 9 kHz is caused by reflections in the cavum conchae back wall. As this HRTF was measured with an artificial head with no ear simulator, the typical resonance in the range of 8 kHz that occur inside the ear canal is not present.

A spherical head-related coordinate system is used (see fig. 2.5) to describe the direction of the sound incidence. The origin of this coordinate system is in the center of the head between the ears. The azimuth angle ϕ rotates counterclockwise between 0° (front direction) and

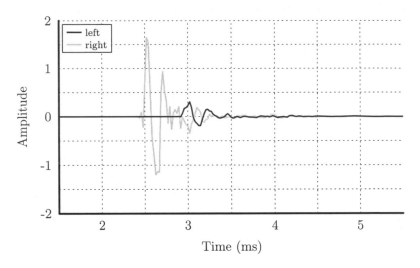

Figure 2.3: Example of head-related impulse response for sound incidence from the right.

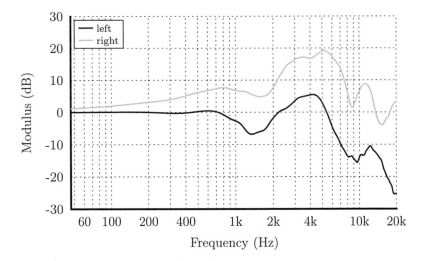

Figure 2.4: Example of head-related transfer function for sound incidence from the right.

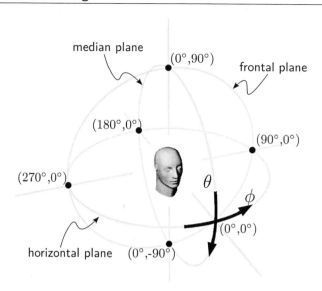

Figure 2.5: Spherical coordinate system used in the HRTF measurements. Elevation angle θ is defined in the range $-90° \leq \theta \leq 90°$ and the azimuthal angle ϕ is defined in the range $0° \leq \phi \leq 360°$.

$360°$. The elevation angle θ is defined from $-90°$ (bottom) to $90°$ (top). Three planes are also defined: 1) the horizontal plane ($\theta=0°$), 2) the frontal plane ($\phi=\pm90°$) and 3) the median (sagittal) plane ($\phi=0°,180°$). Planes parallel to the median plane are called sagittal planes.

Different definitions for the ATF can be found in relevant literature. BLAUERT (1997) describes three definitions of the ATF; one of them will be used throughout this thesis, the free-field HRTF.[5] This describes the transformation from the sound pressure generated by a sound source in far-field and measured at the center of the head (with the head absent) to the pressure generated by the same source at the same position and measured at the entrance of the listener's ear canals.

Another ATF definition used in this thesis is the *directional transfer function* (DTF), which removes the components common to all HRTFs, described by the diffuse-field HRTF (MIDDLEBROOKS, 1999b). The diffuse-field HRTF is computed by taking the root-mean-square (RMS) of the sound pressure at each frequency averaged across all measured HRTFs.

---

[5]Simply called HRTF in the following.

Both the HRTF and the DTF contain, however, no distance information and can thus only be used to render sources at far-field, i.e., plane wave sources. This restriction is usually not a hindrance for virtual reality (VR) applications, especially for room acoustic simulations. However, the most impressive binaural demonstrations occur when sound sources are located in the near-field, increasing the realism of VR scenes (LENTZ, 2007).

A very important recent contribution in the field of individual HRTF is the range extrapolation technique, described by DURAISWAMI et al. (2004) as "a way to obtain the range dependence of the HRTF from existing measurements conducted at a single range!"

2.4.2 Sound Localization

Sound localization is a very complex mechanism performed by the human brain. It is not only dependent on the directional cues contained in the binaural signal captured at the ears, but it is also intertwined with the other senses, especially vision and proprioception (SEEBER, 2002).

While binaural disparities like interaural time and level differences (ITD and ILDs) play an important role for sound localization in the horizontal plane (MACPHERSON and MIDDLEBROOKS, 2002), monaural spectral cues are known to determine the perceived sound-source position in the sagittal planes (top/down, front/back; BLAUERT, 1969). In section 6.2 the localization performance will be discussed with regard to these two mechanisms.

Non-acoustical cues such as head movements, also called sounding (BLAUERT, 1997), are avoided by keeping the listener's head still during the signal presentation.

In fig. 2.6 a new spherical coordinate system is defined that describes the acoustic target's position on the listening experiment, based on a horizontal-polar coordinate system (MORIMOTO and AOKATA, 1984). Again, the origin of the coordinate system is in the center of the head. The lateral angle α is defined from $-90°$ (right) to $90°$ (left). The polar angle β rotates counterclockwise between $0°$ (front) and $360°$. Every lateral angle defines a sagittal plane.

Localization in Horizontal Plane

In the horizontal plane, mainly the binaural cues (ITD and ILD) are used for the localization. The interdependency of these two cues is described by the *duplex theory* (RAYLEIGH, 1907).

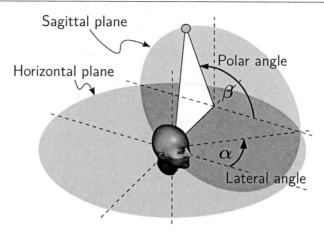

Figure 2.6: Spherical coordinate system used in the localization experiments. Lateral angle α is defined in the range -90° $\leq \alpha \leq$ 90° and the polar angle β is defined in the range 0° $\leq \beta \leq$ 360°.

HARTMANN (1999) summarizes the matter of binaural cues as follows: "The physiology of the binaural system is sensitive to amplitude cues from ILDs at any frequency, but for incident plane waves, ILD cues exist physically only for frequencies above about 500 Hz. They become large and reliable for frequencies above 3000 Hz, making ILD cues most effective at high frequencies. In contrast, the binaural physiology is capable of using phase information from ITD cues only at low frequencies, below about 1500 Hz."

In section 6.2 the localization performance in the horizontal plane is analyzed with regard to the lateral error, which is the RMS of the localization error[6] in the lateral dimension (MIDDLEBROOKS, 1999b).

Localization in the Sagittal Plane

Binaural cues can only assist in the perception of lateral angles on the horizontal plane. Confusion in the sagittal plane, usually experienced as a reversal between front and back, or up and down, may occur as ITD and ILD remain (approximately) constant along the polar angles. The region where such confusions occur is called *cones of confusion* (MILLS, 1972).

---

[6]Localization errors are calculated by subtracting the target angles from the response angles.

The hearing system relies on the monaural cues to assert the polar angle within a cone of confusion. These are direction-dependent spectral coloration of the sound due to the asymmetry of the head and especially the pinna (MIDDLEBROOKS, 1999b).

In section 6.2 the localization performance in the sagittal plane is analyzed with regard to the polar error and the quadrant error (MIDDLE-BROOKS, 1999b). LE is the RMS of the localization error in the polar dimension and is used to quantify the local performance in the polar dimension. QE is the percentage of responses where the absolute polar error exceeded $90°$ and it is used to describe the degree of confusions.

3

Measurement of Individual Head-Related Transfer Function

Depending on the measurement setup used and the amount of directions to be measured, measurements of head-related transfer functions (HRTFs) can be very time-consuming. In contrast to artificial heads that remain static and may therefore be measured over a long period of time, human subjects have difficulty keeping their position—especially their head position—for a longer period of time. Thus, individual HRTF measurements should ideally be conducted in the shortest amount of time possible, to reduce positioning error and improve the comfort of the subject being measured.

A large number of previous work in measuring individual HRTF has been conducted by BRONKHORST (1995); MØLLER et al. (1995a); ALGAZI et al. (2001); ZOTKIN et al. (2006); MAJDAK et al. (2007), and LENTZ (2007), just to name a few. Different setup strategies have been used by the different research groups: e.g. one loudspeaker being moved on an arc with a mechanical stepping device or many loudspeaker fixed on an arc or even several microphones distributed over a sphere. For the setups listed above, typical measurement duration varies from 20 min to 2.5 h to measure 1000 to 1500 HRTFs with an average angular resolution of approximately 5°.

This chapter describes a new HRTF measurement setup that was developed to allow the fast acquisition of individual HRTFs. Moreover, this setup is one of the first of its kind designed to be compatible with the range extrapolation technique. The setup consists of a circular arc and up to 40 broadband loudspeakers that can be distributed (almost) arbitrarily along the arc. By rotating the subject horizontally inside this

---

*Most of the results presented in this chapter have been previously published at

- MASIERO; POLLOW, and FELS (2011a);
- MASIERO; POLLOW; DIETRICH, and FELS (2012);
- DIETRICH; MASIERO, and VORLÄNDER (2012a);
- POLLOW; MASIERO; DIETRICH; FELS, and VORLÄNDER (2012b).

arc, HRTFs are measured at fixed points on a spherical grid, discretized over azimuth and elevation angles. A continuous description of the HRTF can be obtained in a post-processing stage using interpolation algorithms.

This chapter starts by introducing the requirements of the new HRTF measurement setup and describing the hardware solutions chosen to meet these criteria. Then the explanation of the optimized multiple exponential sweep method (MESM) is presented. It is used to accelerate the total measurement duration by over 90% compared with the sequential measurement method and by over 50% compared with the original MESM. Then the post-processing operations of equalization, interpolation, and range extrapolation, which are applied to the measure HRTFs, are discussed.

3.1 Hardware Design

The main aspect common to all modern HRTF measurement setups is that the measurement itself should be concluded in the shortest time possible. A method commonly used in numerical acoustics to reduce computation time is the reciprocity method, where source and receivers are exchanged to be able to simulate multiple receiver points at a single run. An HRTF measurement method based on the principle of reciprocity was proposed by ZOTKIN et al. (2006) using a miniature sound source placed at the entrance of the blocked ear canal and 32 microphones distributed on a spherical array of 0.7 m radius. As the excitation signal has to be played only once for each ear, this results in a very short measurement time. However, the use of a miniature sound source yields a considerably smaller signal-to-noise ratio (SNR) and restricts the measurement frequency range to frequencies above approximately 1 kHz. ZOTKIN et al. (2006) used a model-based extension of generic HRTFs for frequencies below this limit. If a high spatial resolution is desired, many microphones positions have to be measured, which increases the hardware costs, or measurement has to be repeated for different configurations of a smaller array, increasing measurement time.

Microphones can, in contrast to loudspeakers, be miniaturized without severe restrictions to sensitivity levels and working frequency range. A direct HRTF measurement system using two miniature microphones, ideally placed in the entrance of the blocked ear canal (MØLLER et al., 1995b), can provide satisfactory SNR levels. Direct

HRTF measurement setups can be divided into three categories regarding the number and configuration of physical sound sources in the setup.

- **Dense array**: with as many loudspeakers as directions to be measured.

- **Hybrid array**: with a group of loudspeakers placed on an arc, where either the arc or the subject is turned.

- **Sparse array**: only one loudspeaker is moved in every direction to be measured.

If high spatial resolution is desired, a dense setup will require a large number of hardware channels, drastically increasing the costs and complexity of the setup. This is also true for reciprocal measurements. A sparse measurement setup, on the other hand, will always be the slowest of all methods as no parallelization of the measurement procedure is possible. For example, the system built at TNO in the Netherlands needs 2.5 h to complete a measurement with 976 directions (BRONKHORST, 1995). For such setups, it is common that subjects wear a head tracking device to verify that the head position is kept constant throughout the complete measurement duration.

The hybrid array is a trade-off between speed and hardware complexity and is therefore most commonly found, e.g. the setup at the CIPIC Labs in USA that can measure an HRTF set at 1250 directions in approximately 1.5 h (ALGAZI et al., 2001) or the more recently constructed setup from the Acoustics Research Institute (ARI) in Austria that can measure almost 1500 directions in approximately 20 min (MAJDAK et al., 2007). These two setups differ in one main aspect. While in the CIPIC setup the listener sits still and the arc is rotated along the subject's interaural axis, the ARI setup has a static arc and the subject is rotated around its longitudinal axis. Because of the above-mentioned advantages, a hybrid array was chosen for the measurement arc.

Not only the array type varies in between HRTF measurement setups, but also the distance from sound sources to the listener. For example, the CIPIC's system uses an arc with 1 m radius while the system build at the Aalborg University has a radius of 1.95 m (MØLLER et al., 1995a), allowing subjects to be measured in standing position.

This setup should be capable of measuring both near- and far-field HRTF. One possibility would be to construct a number of arcs with different radii, as described by LENTZ (2007). Another possibility is to apply the range extrapolation technique, based on the acoustical spherical holography that enables us to calculate the near-field HRTF from the far-field measurements or vice versa. Therefore, a setup with a single arc and, thus, fixed distance was preferred.

As evanescent spherical waves will have faded away in far-field, a measurement in far-field might not contain enough information to allow the reconstruction of such waves in near-field. For this reason measurements should ideally be conducted as close as possible to the listener's head. On the other hand, to apply acoustical spherical holography, all scattering objects important for the HRTFs (the shoulders and upper torso) must be contained inside of the spherical measurement surface. If measurements were conducted with an arc of short radius, a large region in the bottom of the sampling sphere would remain uncovered, as the listener's body would be on the way of the loudspeakers. Since, POLLOW et al. (2012a) could not verify that outward extrapolation was more robust than inward extrapolation and BRUNGART and RABINOWITZ (1999) stated that "HRTFs are virtually independent of distance for sources beyond 1 m", the measurement arc was planned to have a radius of 1 m.

3.1.1 Sound Source

In acoustical spherical holography, the scattering problem is modeled as a distribution of *point sources* on a sphere containing all the scattering objects (ZOTTER, 2009, p. 34). When placed in the far-field, the radiation pattern of a point source can be approximated by an incident plane wave. Thus, when measuring HRTFs in far-field, regular loudspeakers can be used.

On the other hand, when measuring HRTFs at near-field, these larger loudspeakers are inappropriate as their directivity pattern diverges from that of a point source. Furthermore, two-way loudspeakers are also inadequate because the HRTFs will be blurred as their acoustic center moves from the woofer to the tweeter as the frequencies increase. Attempts have been made to produce acoustic transducers that mimic a point source. BRUNGART and RABINOWITZ (1999) developed a source with an electrodynamic horn driver placed at the end of a long tube. Unfortunately, as this construction acts as a wave guide the frequency response shows many peaks and dips (due to wave interference) and equalization might become problematic. QU et al. (2009) proposed the use of an electric spark as an acoustic point source. Such sources do display a flat frequency response and an almost omnidirectional radiation pattern. However the impulse generated by them is not repeatable and the obtained SNR levels are far below the usual levels obtained using loudspeakers and correlation technique.

An ideal loudspeaker for near-field HRTF measurement setup should meet the following design criteria:

1. Broad-band reproduction

2. Omnidirectional radiation pattern

3. Low distortion artifacts due to nonlinearity

A large membrane is required to be able to generate enough sound pressure at low frequencies. In the other hand, omnidirectional radiation pattern can only be achieved if the driver's membrane is small compared with the wavelength. Thus, a small membrane is required to radiate omnidirectionally at higher frequencies. It is then clear that the choice of a single membrane size will force a trade-off between pressure level at low frequencies and omnidirectionality at high frequencies. The desired frequency range was therefore reduced from the complete audible range to the range between 300 Hz and 20 kHz. This is a reasonable restriction as, according to Møller et al. (1995a), HRTFs show very little individual variation below 300 Hz and its asymptotic behavior can be extrapolated (cf. section 3.3.2). Even with a relaxed frequency range restriction, only a handful of broad-band drivers are still able to meet these requirements.

Three loudspeaker drivers were analyzed regarding their frequency range, maximal sound pressure level (SPL), distortion level, and directivity. The driver with the highest maximal SPL and consequently lowest nonlinear distortion had relatively large dimensions. It was discarded due to its bundled directivity. The other two drivers showed equivalent characteristics, with lower maximal SPL, higher nonlinear distortion, and a smoother directivity pattern in the frontal direction (due to the smaller membrane diameter). Since the loudspeakers on the arc will be relatively close to the microphones, maximal SPL was not defined as a critical parameter and therefore the smaller driver with 32 mm diameter was chosen as it also allowed an easier mechanical fixation at the enclosure

To radiate in the low frequency range the chosen driver needs an enclosure, otherwise the air would just move from the front to the back of the membrane in an acoustic shortcut (reactive intensity) and no sound wave would propagate outwards (active intensity). According to the Thiele-Small parameters calculated for the chosen driver, a volume of at least 100 ml is required to allow a sound reproduction down to 300 Hz. Note that even such a small enclosure can influence the radiated sound field due to its edge diffraction. An optimization of the enclosure was carried out to minimize these effects, as described below.

Figure 3.1: Developed drop-like loudspeaker mounted on an arc element. The perpendicular supporting truss structure allows the loudspeakers to be placed freely within delimited regions of the arc.

Enclosure Optimization

In the first step the driver's membrane velocity was measured with a laser doppler vibrometer at 154 points and these values were used as input data for the loudspeaker simulation. The vibrometry results showed the presence of eigenmodes only at very high frequencies. Three forms of enclosures were simulated: a cylinder with rounded front edge, a cylinder with both front and back edges rounded and a drop-like enclosure. All forms avoid, in varying degrees, sharp edges responsible for diffraction. Simulation results showed that, for a point on axis 1 m away from the membrane, the drop-like enclosure has the least influence in the loudspeakers frequency response and directivity (SARTOR, 2010).

Furthermore, influences due to possible sound reflections by neighboring loudspeakers have to be considered. In order to verify which form yields the best results, another simulation was carried out using three identical loudspeakers placed on an arc with a 1 m radius placed 10° apart. The central loudspeaker was set as the sound source while the other two loudspeakers were left inactive, as mere diffraction bodies. Again, the drop-like enclosure showed a slightly lower influence on the radiated sound field. This form was therefore chosen for the loudspeaker enclosure, as shown in fig. 3.1. The frequency responses of

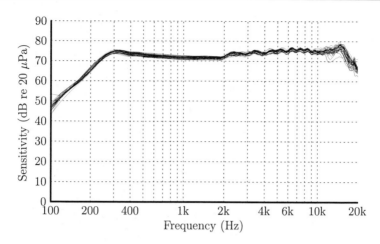

Figure 3.2: Frequency response of 40 drop-like loudspeakers measured at 1 m and 1 V.

the 40 constructed loudspeakers are shown in fig. 3.2. The constructed loudspeakers provide a relatively flat frequency response in the range from 300 Hz to 14 kHz and an acceptable SNR can be achieved up to 20 kHz. The measurement shows low variability between loudspeakers. However, an individual equalization is still required.

3.1.2 Supporting Arc & Head-Rest

The design of the arc that supports the loudspeakers also aims at minimizing the influence of the arc on the radiated sound field to avoid reflection and diffraction effects. Although they are easier to manufacture, bulky structures have a great influence on the sound field and should be avoided. On the other hand, a thin metal rod can be considered acoustically transparent if its diameter is much smaller than the wavelength of the impinging sound wave. The supporting arc was therefore designed with thin metal rods in a trellis structure to minimize disturbing scattering effects while providing sufficient stability.

As mentioned earlier in this chapter, the radius of the supporting arc was defined to be 1 m. As a person has to stay in the middle of the loudspeaker array, the use of a complete circle is not feasible. Hence, an

Figure 3.3: Picture of listener placed inside the developed HRTF measurement system. The metal plate where the listener stands is fixed
on a turntable. The used head-rest can be seen behind the
listener.

arc of 300° was chosen, allowing measurements of elevation angles from
−60° to 90°.[1]

Due to its light construction, the supporting arc displays an under-
damped oscillatory behavior. If the supporting arc where thus to be
rotated, like in ALGAZI et al. (2001), a long settling time would be
necessary until the arc reaches its rest position again. It was decided
to keep the arc stationary and to rotate the subjects inside the arc as
the settling time for the human head is assumed to be shorter than
that of the arc. The subject was rotated with the help of a turntable

---

[1]For more details on the developing stages of the arc, please refer to
(MASIERO et al., 2011a).

	position variation (cm)		
	x	y	z
without head-rest	4.60	5.70	0.30
with head-rest	0.08	0.06	0.10

Table 3.1: Measurement of the head position variation for one person standing still for two minutes with and without a head-rest.

and a head-rest device was used to help the test subjects to keep still during measurement. A controllable turntable, already available at the Institute of Technical Acoustics, was used for this purpose. A head-rest was constructed with a thin metal bar fixed to the turntable at the bottom end and connected at the top end to a Y-shaped metal bar that could be adjusted in elevation and depth to be adapted to the listener's head.

Tests with a subject wearing a position tracking device showed that the natural displacement could be considerably reduced with the help of a read-rest (see table 3.1). This result is only demonstrative as it was carried out with one subject only. During the measurement, it was verified that the most critical aspect of positioning was to place the listener with its longitudinal axis matching the turntable's rotation axis. However, as long as this misplacement remains constant throughout the whole measurement, it might be possible to compensate for it at the post-processing stage (ZIEGELWANGER, 2012).

3.1.3 Data Acquisition and Amplifiers

In order to drive all loudspeakers independently a multi-channel measurement setup was put together. A computer is connected to two multi-channel professional sound cards that are commercially available. These were then connected via an optical interface to five commercial AD/DA converters. The converters' DA output is connected to two 20-channel, low noise and low distortion amplifiers, specially developed for this setup, with a maximum power of 10 W per channel. Two miniature microphones are directly connected to the AD input of one of the AD/DA converters. A connection diagram of the complete system is presented in fig. 3.4.

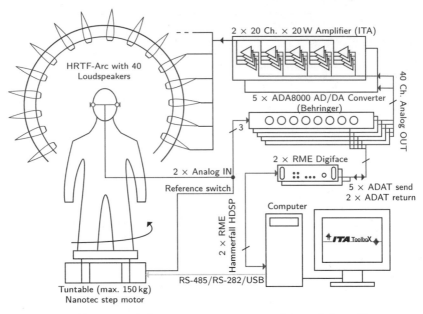

Figure 3.4: Diagram of the measurement setup. Adapted from KRECHEL (2012).

3.1.4 Sampling Grid

The measurement setup described in this chapter allows the loudspeakers to be placed at almost any position on the arc, only limited at the positions where perpendicular truss supporting structure is present (see fig. 3.1). As a hybrid model was chosen, where the listener is rotated in azimuth and the loudspeakers have a fixed elevation, only axisymmetric grids can be used. To improve the measurement time, the chosen grid should measure at each azimuth position as many points in elevations as possible. Examples of such spatial grids are the equiangular grid, the Gaussian grid, and the IGLOO grid (ZHANG et al., 2012).

A longitude-latitude grid is any grid where points are equally distributed in T elevation angles ($\Delta\theta = 180°/T$) and U azimuth angles ($\Delta\phi = 360°/U$). An equiangular grid is formed when $\Delta\theta = \Delta\phi$. For calculations in the SH-domain (spherical harmonic domain, see section 3.3.2), the most efficient longitude-latitude grid is the Gaussian grid, whose elevation angles are defined at the roots of the Legendre polynomial for the desired order and which the inverse problem can be efficiently solved using the weighted Hermitian of the matrix containing the spherical harmonic base functions (ZOTTER, 2009).

These kind of sampling schemes display a concentration of points at the poles ($\theta = 0°$ and $\theta = 180°$). To avoid this concentration and still allow a similar interpolation quality, the IGLOO method discards points in the polar region (ZHANG et al., 2012), therewith achieving a greater sampling efficiency,[2] though losing the advantage of efficient calculation in the SH-domain.

The placement of the loudspeakers in the constructed setup shows some minor deviations from the exact positions of the desired sampling scheme. These deviations are caused by the structural restrictions previously mentioned. Therefore, the exact locations of the loudspeakers have to be determined so that they can be used for further data processing. A method to extract the loudspeakers' position based on acoustic measurements was described by KRECHEL (2012). A system consisting of two microphones mounted on an aluminum bar placed exactly 40 cm from each other is therefore used. This bar is mounted on a support arm, which is fixated at a stand, itself attached to the center of the turntable.

The transfer function between loudspeakers and the two microphones are measured for at least two azimuthal positions. The travel time τ_{ikj} between the ith loudspeaker to the kth microphone at the jth measurement position is estimated by first convolving the measured impulse response with its minimum-phase version, then taking the Hilbert transform of the resulting convolution and finally searching for the zero-crossing in the Hilbert-transformed signal closest to the maximum value of the original convolution result.

A system of linear equations is set up using the estimated distance between the microphones and the known angle of the turntable during the measurements to obtain the exact position of the loudspeakers.

$$x_{i,\text{opt}} = \arg\min_{x_i} \sqrt{\sum_{k=1}^{2} \sum_{j=1}^{n} \sum_{i=1}^{L} \left[\|x_i - P_{kj}\| - \left(\tau_{ikj} - \frac{\delta_{\lceil i/8 \rceil}}{f_{\text{s}}} \right) \cdot c \right]^2}$$

(3.1)

where $\delta_{\lceil i/8 \rceil}$ is the latency of each sound card,[3] and c the speed of sound. x_i is the position of the loudspeaker and P_{kj} is the position of the kth

---

[2] ZOTTER (2010) defines the *sampling efficiency* as the ratio between the number of points in the actual grid and the minimum number of points required to correctly represent the spherical harmonic order O, which is the maximum order obtainable with the array without spatial aliasing (please see section 3.3.2 for more details).

[3] Each sound card has eight channels and it is assumed that all channels have the same latency.

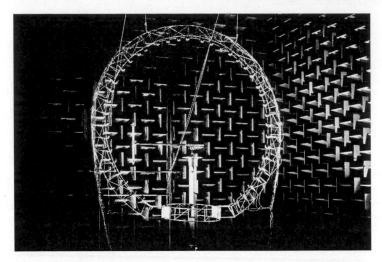

Figure 3.5: Setup developed for the measurement of the real position of the loudspeakers. Two microphones are placed at the top and bottom of the vertical bar at an exact distance of 40 cm from each other. The bar is fixated by a stand to the turntable and is turned to provide multiple measurement positions.

microphone at the jth measurement position, given by equation

$$P_{kj} = \begin{pmatrix} x \\ y \\ z \end{pmatrix} = \tag{3.2}$$

$$\begin{pmatrix} \sin(\gamma) \cdot r_k \cdot \cos(\phi_j) - [\sin(\rho) \cdot d + \cos(\rho) \cdot \cos(\gamma) \cdot r_k] \cdot \sin(\phi_j) \\ \sin(\gamma) \cdot r_k \cdot \sin(\phi_j) + [\sin(\rho) \cdot d + \cos(\rho) \cdot \cos(\gamma) \cdot r_k] \cdot \cos(\phi_j) \\ - \cos(\rho) \cdot d + \sin(\rho) \cdot \cos(\gamma) \cdot r_k \end{pmatrix},$$

where r_k is the distance between the kth microphone and the center of the bar holding the microphones, ϕ_j is the angle of the turntable at the jth measurement position, d is the distance and ρ the angle between the bar holding the microphones and the stand connected to the turntable, and γ is the torsion of the bar hold the microphones. The fact that the obtained system of equations is overdetermined is very practical as it allows not only to estimate the position of the loudspeaker, but also the orientation of the bar, its distance to the rotation axis, the latency of each sound card, and the speed of sound at the time of measurement.

3.2 Excitation Signal

The maximum speedup in the measurement time is achieved if all loud-speakers on the arc can be used simultaneously during the measurement. XIANG and SCHROEDER (2003) and VANDERKOOY (2010) showed that parallel measurement can be performed with pseudo-random sequences, since they are mutually orthogonal. However, just as in the single channel case, measurements carried out using pseudo-random sequences are very sensitive to time-variance and nonlinearity in the measurement chain (MÜLLER and MASSARANI, 2001). As discussed in section 2.3, sweeps offer greater robustness for measuring systems that do not fully comply with the LTI assumption, e.g. weak time variances (slow changes of the system response over time) or nonlinear transfer characteristics (harmonic distortion). Furthermore, with a little tweak, sweeps can also be used for a multiple parallel excitation.

Instead of mutual orthogonality, the LTI principle of superposition is now analyzed: if the system output $g(t)$ is composed by the addition of two or more filtered versions of the (time-shifted) input signal $s(t)$,

$$g(t) = \sum_i h_i(t) * s(t - \tau_i), \tag{3.3}$$

then the deconvolution of $g(t)$ by $s(t)$ will result in

$$h'(t) = \sum_i h_i(t - \tau_i). \tag{3.4}$$

As long as the input signals are adequately shifted in time, the IRs $h_i(t)$ can still be restored from $h'(t)$. MAJDAK et al. (2007) introduced a new fast measurement method for weakly nonlinear systems by using exponential sweeps and an optimization strategy to overcome the interference in the measurement between nonlinearities—appearing as noncausal harmonic IRs—and the system's IRs. Two different strategies to avoid this interference were proposed and combined using an optimization algorithm, with respect to either measurement time or SNR, yielding the so called *multiple exponential sweep method* (MESM).

It will be shown that, if the reverberation time of the room where the measurement is conducted is small enough, then a generalized over-lapping strategy is sufficient (DIETRICH et al., 2012a). Furthermore, assuming that only the first 5 ms of the measured IR contains important information for the HRIR—the rest being unwanted reflections—the sweeps can be overlapped even closer to each other, yielding an even faster measurements with unchanged accuracy. Using an optimized version of the multiple exponential sweep technique, approximately four thousands discrete points can be measured within less than six minutes.

3.2.1 Multiple Exponential Sweep Method

The MESM proposed by MAJDAK et al. (2007) reduces the measurement duration significantly compared with sequential measurements using the exponential sweep method where the number of loudspeakers is high. Using the traditional sequential method (SM), the duration of a measurement made with N loudspeakers is given by

$$T_{\text{SM}}(N) = N \cdot (\tau_{\text{sw}} + \tau_{\text{st}}), \tag{3.5}$$

where τ_{sw} is the length of the excitation sweep and τ_{st} is the stop margin, i.e. the time required to allow the system to decay after the sweep has ended.

Using the MESM, the sweeps are played back with a certain waiting time or delay τ_{w} between each subsequent sweep. Hence, sweeps of several loudspeakers might run (partly) in parallel. As a new sweep starts every τ_{w}, in the ideal case without any nonlinearities, the measurement duration with N loudspeakers is given by the sum of the waiting time of the first $N - 1$ sweeps plus the length of the last sweep τ_{sw} and the required stop margin τ_{st}, thus

$$T_{\text{MESM}}(N) = (N - 1)\tau_{\text{w}} + \tau_{\text{sw}} + \tau_{\text{st}}. \tag{3.6}$$

Usually the length of excitation sweeps used for HRTF measurement lies in the range from 0.2 s to 2 s, for very short to moderately long sweeps. If the measurements are conducted in suitable anechoic environments, τ_{w} is estimated to be in the range from 20 ms to 200 ms and $\tau_{\text{st}} = \tau_{\text{w}}$. Comparing eq. (3.5) with (3.6) and using the nominal values listed above, a theoretical speedup of 88% can be expected with the MESM when compared with the SM. The parallel measurement shows great potential for a large N, long sweeps, and a short waiting time. Hence, the minimization of this delay is of interest.

Again, in the ideal case without any nonlinearity, the smallest possible value for τ_{w} is the reverberation time τ_{IR} of the room where the measurements are conducted, as described in fig. 3.6. However, if the system is weakly nonlinear, the noncausal harmonic IRs could be superposed

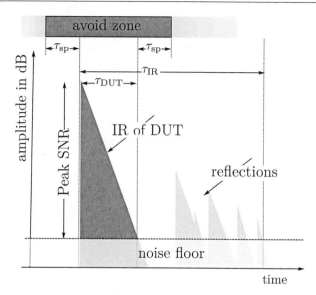

Figure 3.6: Temporal structure of an IR of a linear system measured in an anechoic environment with an exponential sweep. Reflections are caused by the measurement setup itself and other foreign objects in the room.

with the IR of interest, irrevocably corrupting the measurement. MAJ-DAK et al. (2007) suggested the use of *overlapping* (OL) and *interleaving* (IL) to avoid this from happening (see fig. 3.7).

When overlapping the harmonic IRs appear between the IRs of interest as shown in fig. 3.7(a). A drawback of the occurrence of harmonic IRs is the waiting time $\tau_{w,OL}$ that has to be increased (compared to the ideal situation)so that it does not interfere with the region of interest. Furthermore, the sweep rate can be increased to shorten the delay between fundamental and harmonic IRs. The maximum order of harmonics k_{max} present in the measurement has to be finite, and preferably small, to allow a small $\tau_{w,OL}$.

When interleaving, η IRs of interest are grouped together, placing as many fundamental IRs as possible in the time span between the first fundamental IR and its corresponding first harmonic IR, as illustrated in fig. 3.7(b). Contrary to overlapping, the sweep rate in this case should be decreased to enlarge the delay between fundamental and harmonics and thus fit more IRs of interest (i.e. fundamental IR) inside this gap.

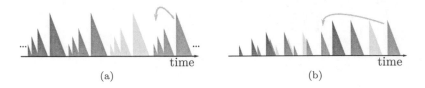

Figure 3.7: Schematic example of the temporal structure of the measurement of a weakly nonlinear system limited to three harmonic
IRs obtained with (a) the overlapped IR method and (b) the
interleaved IR method (with $\eta = 4$).

MAJDAK et al. (2007) also described how to combine both methods
using the two different time delays $\tau_{w,IL}$ and $\tau_{w,OL}$. These methods
overlap groups of interleaved sweeps, as illustrated in fig. 3.8(a). They
describe that the optimum solution minimizing measurement duration
is given by the interleaving waiting time

$$\tau_{w,IL} = \tau_{IR}, \tag{3.7}$$

that depends only on τ_{IR} and the overlapping waiting time

$$\tau_{w,OL} = \Delta t_k + \eta\,\tau_{IR} \tag{3.8}$$

where Δt_k is the time interval between the desired IR and the furthest
harmonic IR (cf. fig. 2.2), which gives the time distance between the
beginning of the first harmonic IR and the end of the last desired IR
belonging to the same interleaved block. DIETRICH et al. (2012a) showed
that

$$\Delta t_k = \frac{\log_2(k_{max})}{r_{sw}}, \tag{3.9}$$

where r_{sw} is the sweep rate.

According to WEINZIERL et al. (2009), the optimum value for η is
given by

$$\eta = \left\lceil \frac{\tau_{IR} - \tau_{IR,2} + {}^{1}/r_{sw}}{\tau_{IR}} \right\rceil, \tag{3.10}$$

where $\tau_{IR,k}$ is the length of the kth harmonic RIR (see section 2.3).

The SNR and the temporal and spectral structure of the results
obtained with sequential measurements and the MESM will remain the
same if the following requirements are met.

1. The system is weakly nonlinear, i.e. the number of harmonic IRs
   present in the measurement is small.

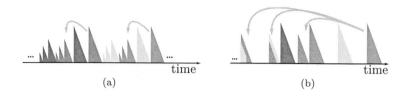

Figure 3.8: Schematic example of an IR measured with (a) the MESM as suggested by MAJDAK et al. (2007) with $\eta = 2$ and (b) with the optimized MESM described in this work. Note that no harmonic IR superposes the desired fundamental IRs.

2. In case nonlinearity is observed, the output level must be kept constant during the actual measurement and the calibration measurement (used to determine the number of harmonic IRs and actual length of the desired IR).

3. The smallest delay τ_w between two subsequent sweeps must be larger than the length of the desired IRs.

4. nonlinearity should be restricted to elements of the measurement chain where the excitation signal is not yet superposed, i.e. amplifiers and loudspeakers. Microphones and preamplifiers have to be driven in their linear range only as nonlinearity at this stage will introduce inter-modulation that might corrupt the measurement if not taken into account at the optimization stage.

3.2.2 Optimized MESM

MAJDAK et al. (2007) claim that their MESM provides minimal measurement duration. However, while analyzing the temporal structure of a usual HRIR measured in an anechoic environment, DIETRICH observed that an additional speedup of the measurements was still possible.[4] As will be shown later in this section, if the region of interest is only a small fraction of the measured IR, a generalized overlapping strategy (with wait time $\tau_{w,OPT}$) can further accelerate the measurement process.

Temporal Structure of Measured IR

It is sensible to assume that the IR of an acoustic system is causal and that its energy decays exponentially. The measurement of an HRTF is

---

[4]Personal communication with Pascal Dietrich in 2011.

equivalent to the measurement of directional transfer functions, where the *device under test* (DUT) is the listener's head and loudspeakers are used as sound sources.

In such measurements, the obtained IRs consist of a direct sound path plus reflection and diffraction of the listener's body, containing the desired spectral and directional information, followed by reflections due to objects or room boundaries which can be understood as unwanted artifacts. This situation is depicted in fig. 3.6. The overall length of these IRs τ_{IR} is limited by the moment all reflections cease or disappear below background noise. These reflections might still occur even in *anechoic* environment, caused either by a hard floor (hemi-anechoic chamber) or by other necessary objects in the room, e.g. lamps, support frames, doors, pedestals, etc.

It is important to notice that only the beginning of the IR, with the length τ_{DUT}, has to be protected against reflections and harmonic IRs—that might overlay the desired IR during a measurement with MESM (DIETRICH et al., 2012b). Hence, an *avoid zone* around the IR of the DUT is defined by adding a safety region τ_{sp} before and after the desired response.

The definition of the avoid zone suggests that, in contrast to the MESM, the regions containing only unwanted reflections can be used to place the harmonic IRs. The overlapping method, and hence the MESM, can directly benefit from this observation by adapting eq. (3.8) to

$$\tau_{\text{w,OL}} = \frac{\log_2(k_{\text{max}})}{r_{\text{sw}}} + (\eta - 1)\,\tau_{\text{IR}} + \tau_{\text{DUT}} + \tau_{\text{sp}}. \qquad (3.11)$$

Equations (3.7) and (3.10) remain unchanged, resulting in a shorter measurement duration.

Placement Strategies for Harmonic IRs

The harmonic IRs present in the measured IR do not necessarily have to be cumulated in blocks, as advocated by the MESM (DIETRICH et al., 2012b). The only constraint for a valid measurement is that no harmonic IRs fall into the avoid zones, as illustrated in fig. 3.8(b).

As a practical consideration, the measurement system described in this thesis is only weakly nonlinear and can be reasonably quantified by claiming a value for total harmonic distortion below 10% for all frequencies.[5] This results in an attenuation of all harmonics of at least 20 dB. As harmonics show the same decay rates as the fundamental IR, harmonics $\tau_{\mathrm{IR,k}}$ will always be shorter than the fundamental IR τ_{IR}. The maximum order $k_{\max}$, before the harmonics fall below background noise. The length of each harmonic IR can be obtained from a calibration measurement using sequential sweep measurements or it can be estimated from

$$\tau_{\mathrm{IR,k}} = \frac{\mathrm{SNR} - a_k}{\mathrm{SNR}} \tau_{\mathrm{IR}}, \tag{3.12}$$

where a_k is the minimum difference (in dB) between the spectrum of a harmonic k compared with the spectrum of the fundamental.

To avoid the desired IR to be corrupted by the room reflections present in the previous IR, the waiting time between sweeps must fulfill $\tau_{\mathrm{w}} \geq \tau_{\mathrm{IR}}$. Considering that all harmonic IRs must fit between two subsequent desired fundamental IRs, the waiting time constraint must be extended to satisfy

$$\tau_{\mathrm{w}} \geq \max\left(\tau_{\mathrm{DUT}} + 2\,\tau_{\mathrm{sp}} + \max(\tau_{\mathrm{IR,k}}), \tau_{\mathrm{IR}}\right) \tag{3.13}$$

Additionally, the start of each kth harmonic IR must fall after the end of an avoid zone and its end must appear before the next avoid zone starts. Both constraints can be written as

$$(\Delta t_k \bmod \tau_{\mathrm{w}}) \geq \tau_{\mathrm{DUT}} + \tau_{\mathrm{sp}} \tag{3.14}$$

and

$$(\Delta t_k \bmod \tau_{\mathrm{w}}) + \tau_{\mathrm{IR,k}} \leq \tau_{\mathrm{w}} - \tau_{\mathrm{sp}}. \tag{3.15}$$

Combining the above-mentioned constraints and also substituting eq. (3.9) results in

$$\tau_{\mathrm{DUT}} + \tau_{\mathrm{sp}} \leq \left(\frac{\log_2(k)}{r_{\mathrm{sw}}} \bmod \tau_{\mathrm{w}}\right) \leq \tau_{\mathrm{w}} - \tau_{\mathrm{sp}} - \tau_{\mathrm{IR,k}}. \tag{3.16}$$

Optimization of Parameters

No analytic solution is known for finding the values $(\tau_{\mathrm{w}}, r_{\mathrm{sw}})$ that satisfy the inequalities (3.13) and (3.16) while minimizing τ_{w} and thus the

---

[5]Loudspeakers commonly present higher distortion levels at lower frequencies. The use of a shelving filter to suppress power at this region can reduce nonlinearity and consequently reduce the size of the harmonic IRs with the consequence of decreasing the observed SNR at this frequency range.

measurement's duration. The straightforward approach to solve this problem is to use exhaustive search (DIETRICH et al., 2012a). The two-dimensional search space (r_{sw}, τ_w) can be normalized by the τ_{IR}. The resulting normalized search space is $(r_{sw} \cdot \tau_{IR}, {}^{\tau_w}/\tau_{IR})$ and it has the advantage that the optimization procedure becomes independent of the IR's length. The normalized size of the avoid zone is then given by

$$\alpha = \frac{\tau_{DUT} + 2\tau_{sp}}{\tau_{IR}}. \tag{3.17}$$

It will be later shown that the smaller the value of α, the higher the chance that the new method will yield a faster measurement than the MESM. The solution is, however, dependent on the parameters k_{max} and $\tau_{IR,k}$.

Valid combinations of (τ_w, r_{sw}) are shown in fig. 3.9 for an example with $k_{max} = 4$, $\tau_{sp} = 0\,s$, $\alpha = 1$ and no decrease of the length of the harmonics: $\tau_{IR,k} = \tau_{IR}$. Valid combinations can always be found for high sweep rates and long delays as in this region the method is equivalent to the original overlapping method. For moderate r_{sw} the allowed τ_w is shorter than the overlapping method (DIETRICH et al., 2012b). For very low sweep rates, the range of valid delays resulting in valid solutions becomes increasingly smaller, so that almost no valid stable solutions can be found with the numeric search algorithm used due to a finite number of discrete search points. Because of its instability, the optimized MESM should be avoided for very small r_{sw}.

Due to a strong fluctuation of the minimum delay over the sweep rate observed in fig. 3.9, it becomes evident that the sweep rate should not be fixed prior to the search. A better approach is to define a search range for the r_{sw} rate and choose the r_{sw} corresponding to the minimum value of τ_w. The change in SNR caused by varying r_{sw} can be neglected in most cases.

3.2.3 Numerical Comparison

As the original MESM does not take into account the temporal structure of the IR, the optimized method is compared with the original method with $\tau_{IR} = \tau_{DUT}$. For comparison, the maximum number of harmonics is set to $k_{max} = 4$ and $\tau_{IR,k} = \tau_{IR}$, which can be seen as a worst case scenario for typical loudspeakers. As displayed in fig. 3.10, both methods always result in a minimum normalized delay shorter or equal to the delay obtained with just the overlapping method. The new method

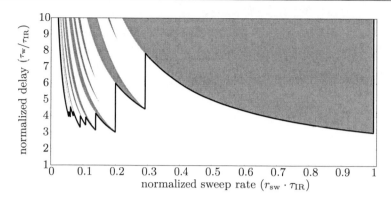

Figure 3.9: Normalized possible solutions for $k_{max} = 4$, $\alpha = 1$, $\tau_{IR,k} = \tau_{IR}$ (white: interference of harmonics with fundamentals, gray: no interference, black: minimum possible delay between sweeps)

shows slightly lower values only for sweep rates in the lower-mid ranges and for low values of α.

To conclude, a comparison is made for real values obtained using the HRTF measurement setup described in section 3.1 placed in the hemi-anechoic chamber in ITA. The desired HRIRs are very short, with an approximate duration of $\tau_{HRIR} = 4\,\text{ms}$ (HAMMERSHØI and MØLLER, 2005). On the other hand, the IR of the hemi-anechoic chamber (containing reflections from the floor, supports, mounts, and doors) has a length in the order of $\tau_{IR} = 40\,\text{ms}$, thus $\alpha = 0.1$.

Sufficient SNR can be achieved with $\tau_{sw} = 1.5\,\text{s}$, which yields over 80 dB peak-to-noise ratio for the desired HRIR. The frequency range of interest is defined from 0.1 to 18 kHz. This corresponds to a sweep rate of $r_{sw} \approx 5$. The avoid zone was enlarged by $\tau_{sp} = 1\,\text{ms}$. The maximum observed harmonic order was $k_{max} = 5$ and a_k was defined in the frequency-domain: $a_2 = -35\,\text{dB}$, $a_3 = -45\,\text{dB}$, $a_4 = -40\,\text{dB}$ and $a_5 = -40\,\text{dB}$.

The best combination of sweep rate and delay found with the optimization algorithm in the region around $r_{sw} = 5$ was $r_{sw,opt} = 5.59$ and $\tau_{w,opt} = 48.095\,\text{ms}$. The new sweep has a length of 1.34 s and the theoretical change in SNR caused by the shorter excitation signal is estimated to be $\Delta\text{SNR} = 10\log_{10}\left(^{r_{s,opt}}/r_{sw}\right) = -0.48\,\text{dB}$.

The sequential measurement of $N = 40$ loudspeakers will take 53.71 s to conclude. Using the original MESM proposed by MAJDAK et al. (2007), the measurement time is reduced to 7.39 s with an average waiting time

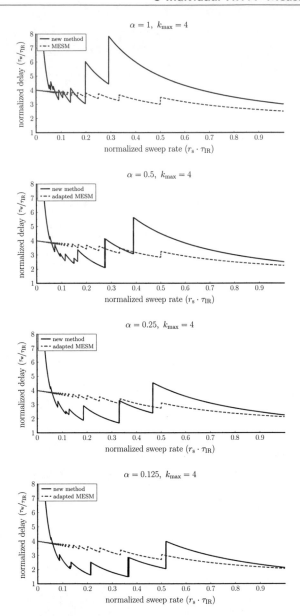

Figure 3.10: Comparison of minimum normalized delay obtained with the original and the optimized MESM for different values of α and $k_{\max} = 4$.

Method	Duration
Sequential Measurement	90 min 01 s
MESM	12 min 49 s
MESM (after eq. (3.11))	12 min 21 s
Optimized MESM (*new*)	5 min 52 s

Table 3.2: Example of the overall measurement duration for a grid with 40 positions in elevation and 100 positions in azimuth, a sweep length of 1.34 s and assuming the turn table takes 0.3 s to reach its next position.

$\bar{\tau}_{\text{w,MESM}} = 155.1$ ms. Taking the time structure of the measured IR in consideration, as suggested in section 3.2.2, can further reduce the measurement time to 7.11 s with $\bar{\tau}_{\text{w,MESM}} = 147.9$ ms. Finally, the optimized MESM described in this thesis will bring the measurement time down to 3.22 s.

When comparing the results obtained using all four methods in the frequency-domain, a maximum deviation of ± 0.1 dB over the entire frequency range of interest can be observed. These deviations are within the repeatability variation observed when measuring the same object with the same measurement method after reposition. Hence, the new method—in the same manner as the MESM—does not introduce noticeable errors if the previously introduced requirements are met.

The measurement duration with the newly proposed method can therefore be reduced to 6% of the time required for the sequential method. This factor can be improved when more channels are interleaved. The theoretical limit for the maximum achievable reduction is estimated from

$$\lim_{L \to \infty} \frac{T_{\text{MESM}}(L)}{T_{\text{ES}}(L)} = \lim_{L \to \infty} \frac{(L-1)\tau_{\text{w}} + \tau_{\text{sw}} + \tau_{\text{st}}}{L\left(\tau_{\text{sw}} + \tau_{\text{st}}\right)} = \frac{\tau_{\text{w}}}{\tau_{\text{sw}} + \tau_{\text{st}}}. \quad (3.18)$$

The reduction would then reach 3.6 % for the parameters used in this example. The total measurement time will depend on the total number of azimuth and elevation directions that should be measured and the time it takes for the turntable to move from one to the next azimuth position. The overall measurement duration for a reasonable[6] spatial resolution with 40 positions in elevation and 100 positions in azimuth and assuming the turntable takes 0.3 s to reach its next position is given in table 3.2.

---

[6] In accordance with the resolution suggested by ZHANG et al. (2012).

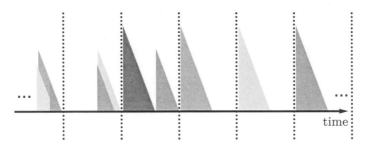

Figure 3.11: Schematic example of an IR measured with the optimized
MESM described in this thesis. The dotted vertical lines represent the limits where each signal is cropped.

3.3 Post-Processing

When measuring the HRIR using any of the MESM described in the
previous section, the measurement's raw output is a series of overlapped
IRs, as illustrated in fig. 3.8(b). It is then necessary to extract each
direction's IRs out of the raw IR. Knowing the delay time τ_w used to
generate the excitation signal, the raw IR is then cropped in L signals,
each has the length τ_w, as exemplified in fig. 3.11. Note that the first
dotted line represents the instant $t = 0$ and the harmonic IRs on the
left of it will actually appear at the end of the raw IR due to the wrap
around effect (see section 2.1).

As described in section 2.3, one of the side effects of the regularized
deconvolution is the occurrence of pre-ringing prior to the actual IR.
Even though these ripples are constituted only by frequencies outside the
desired frequency range, windowing them out might cause more harm
than good. Therefore, first a minimum-phase regularized deconvolution
is performed guaranteeing that no pre-ringing is introduced by the
regularized spectral inversion. After cropping the IRs, each directional
IR is further all-pass filtered by $A_{reg,AP}$ (described in section 2.3) to
extract the effects of minimum-phase regularization.

The cropped IRs contain the desired HRIR plus room reflections
and harmonic IRs from other channels. These unwanted components
can be discarded by time windowing, as shown in fig. 3.12.

The start times of the IRs vary as they depend on the direction of
incidence of the sound. Thus, care should be taken as to where to set the
time window to prevent the desired part of the IR from extrapolating

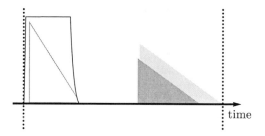

Figure 3.12: Schematic example of how unwanted room reflections and harmonic IRs are windowed out of the cropped IR.

the limits of the window. A straightforward approach is to identify the beginning of each IR using a peak detection algorithm, as for example the one described in the standard *"ISO 3382 – Measurement of room acoustic parameters – Part 1: Performance spaces"*, and to set the time window to start a few samples before the start of the actual IR. Such methods are, however, prone to uncertainty caused by noise, especially in case of contralateral HRIRs, where the SNR is intrinsically low. ZIEGELWANGER (2012) analyzed this effect and developed a model to robustly estimate the start time of the IRs in each direction of incidence, even for situations where the head was not adequately centered during measurement, as already commented in section 3.1.2. The length of the time window should also be adequately chosen so that all unwanted reflections are discarded.

3.3.1 Equalization

The HRIR obtained at this point is free of the influence of measurement artifact. However, it still contains the influence of the loudspeaker's and microphone's frequency response. The most common method to eliminate these influences is the *free-field equalization*, where each HRTF H is divided by the reference transfer function H_{ref}, defined as the transfer function between the same sound source placed at the same position of the HRTF measurement and the same microphone used in the measurement placed at the point corresponding to the center of the head while the subject is not present. This equalization results in the *free-field HRTF* defined by BLAUERT (1997) as

$$H_{\mathrm{free\text{-}field}}(\theta, \phi, r, f) = \frac{H(\theta, \phi, r, f)}{H_{\mathrm{ref}}(r, f)}. \tag{3.19}$$

There are, however, some practical aspects that should be observed while calculating the free-field HRTF. Like the HRIR, the reference transfer function should also have the room reflections windowed out, otherwise undesirable artifacts will occur since reflection paths are not identical in both situations and do not cancel out. Furthermore, care must be taken as the procedure described above will render the resulting HRTF noncausal for the ipsilateral directions,[7] which can have disastrous consequences for the naïve signal processing. Thus, the IRs are cropped, keeping only the range where the window was applied and the time t^i_{win} where the window starts. The cropped IR segments are then divided. As both segments contain approximately the same (small) delay, the resulting IR is not expected to be noncausal.

At this point, the low-frequency asymptote correction is applied. HAMMERSHØI and MØLLER (2005) argues that at low frequencies the human head ceases to act as a scattering object for the incident sound wave, so that the ratio between the transfer function to the ears and to the reference microphone tends to 1. Acoustic data acquisition software usually disregards frequencies close to 0 Hz, nevertheless, HAMMERSHØI and MØLLER (2005) show that when the interpolation is conducted by simply padding the HRIR with trailing zeros, an erroneous value at 0 Hz can have a strong influence up to the mid-frequency range. The correction is applied by simply substituting the value corresponding to 0 Hz by 1.

The delay removed prior to cropping must be reinstated. Therefore, the signal should be padded with zeros to provide a proper time shift. To avoid a sharp transition between the IR and the padded zeros, a fade-in/fade-out operation should be applied to the limits of the cropped IR. After the fading, zeros are padded to the cropped signal and subsequently, the signal is shifted by $t^i_{\text{win}} - t^{\text{ref}}_{\text{win}}$, resulting in the free-field HRTF, or respectively, free-field HRIR.

To apply the range extrapolation, a last step is required. The free-field HRTF should be multiplied by the transfer function between a point source placed at the acoustic center of the loudspeaker to an ideal receiver placed at the acoustic center of the reference microphone. The distance between these two ideal transducers can be estimated from the measured reference transfer function.

Furthermore, if the directional transfer function (DTF) is desired, the diffuse-field HRTF should be estimated from the available free-field

---

[7]The ipsilateral HRIRs are noncausal because the acoustic path from the source to the ipsilateral ear is shorter than the path from the source to the reference microphone.

HRTF at directions $\boldsymbol{\theta}_k$ by

$$H_{\text{diff}}(f) = \sqrt{\sum_k w_k |H(\boldsymbol{\theta}_k, f)|^2}\,, \qquad (3.20)$$

where the weights w_k depend on the sampling grid that is used (DRISCOLL and HEALY, 1994). According to MIDDLEBROOKS (1999a), the DTF is afterwards obtained by dividing the free-field HRTFs by the minimum-phase spectrum of $H_{\text{diff}}(f)$.

In an optional post-processing step the measured data can be smoothed in frequency-domain. It was shown that by smoothing the HRTFs (or DTFs) up to a certain degree, the localization accuracy will not deteriorate (KULKARNI and COLBURN, 1998; BREEBAART and KOHLRAUSCH, 2001; XIE and ZHANG, 2010). A spatial interpolation might also introduce a *spatial smoothing*. To the best of the author's knowledge, the psycho-acoustical effect of the HRTF spatial smoothing has not been studied yet.

3.3.2 Interpolation

As discussed in section 3.1.4, the HRTFs are measured at discrete points distributed on a spherical surface centered on the listener's head. Virtual reality applications require a smooth directional transition when synthesizing moving sources. To directly switch between neighboring HRTFs without any audible artifact, the angular distance between these HRTFs should be smaller than the minimum audible angular difference perceived by the human auditory system—depending on signal type and direction, as low as 1° (BLAUERT, 1997)—a very dense sampling grid would be necessary. This would require a more complex measurement setup and large data storage. Another problem that could occur is that the acoustic simulation software that is used to process the HRTFs cannot handle the sampling grid used for the measurement. A solution for both situations is to interpolate the missing HRTFs from the original data set.

POLLOW et al. (2012a) divide the HRTF interpolation methods into two categories. Local interpolation methods use only the immediate neighboring HRTFs for calculations (LANGENDIJK and BRONKHORST, 2000; FREELAND et al., 2007; LENTZ, 2007) while global interpolation methods use the entire HRTF set for the interpolation (KISTLER and WIGHTMAN, 1992; EVANS et al., 1998; DURAISWAMI et al. , 2004; WANG et al., 2009). The first category has the advantage that its

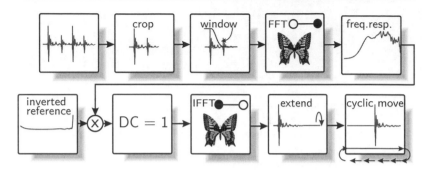

Figure 3.13: Diagram describing the post-processing stages applied to obtain a free-field equalized HRTF from the raw measurement.

calculation can be conducted in a very fast manner. However, HRTFs interpolated with the latter group tend to have a better agreement with measured data. Global interpolation methods also perform a spatial smoothing on the processed data, thus providing robustness against additive noise up to a certain level.

The two main mathematical tools used for a global interpolation in the sphere are the *principal component analysis* (PCA) and the *spherical harmonic* (SH) approximation. EVANS et al. (1998) concluded that the latter method provided better interpolation results than the former technique. Moreover, DURAISWAMI et al. (2004) showed that using the concept of spherical holography, which in turn is based on the spherical harmonic decomposition, one can not only interpolate the HRTF over a spherical surface, but can also extrapolate its radial dependence, an advantage that no other interpolation technique can offer.

The spherical harmonics define a set of orthonormal basis over the spherical surface. They are defined (for the coordinate system defined in fig. 2.5) as

$$Y_n^m(\phi, \theta) \equiv \sqrt{\frac{(2n+1)}{4\pi} \frac{(n-m)!}{(n+m)!}} \cdot P_n^m(\sin\theta)e^{jm\phi}, \qquad (3.21)$$

where P_n^m are the Legendre functions of order n and degree m. Any arbitrary function $p(\phi, \theta)$ defined on a sphere can then be expanded as

$$p(\phi, \theta) = \sum_{n=0}^{\infty} \sum_{m=-n}^{n} c_{nm} Y_n^m(\phi, \theta), \qquad (3.22)$$

where c_{nm} are complex spherical expansion coefficients (WILLIAMS, 1999). These coefficients can be obtained from

$$c_{nm} = \oint_S p(\phi, \theta) Y_n^m(\phi, \theta)^* \, dS. \tag{3.23}$$

Like the continuous Fourier transform and its discrete counterpart, the DFT, spherical harmonics can also be sampled at a grid of points and still correctly describe the whole space, provided that a suitable sampling grid was used and that the data is *spatially* band limited (ZOTTER, 2009, ch. 4). In this case, eq. (3.23) can no longer be applied. The constant c_{nm} must be estimated from a system of linear equations composed of the equation eq. (3.22) evaluated at every sampled direction. This can be cast in a matrix form as

$$\begin{bmatrix} p_1 \\ p_2 \\ \vdots \\ p_k \end{bmatrix} = \begin{bmatrix} Y_1(\phi_1, \theta_1) & \cdots & Y_1(\phi_k, \theta_k) \\ Y_2(\phi_1, \theta_1) & \cdots & Y_2(\phi_k, \theta_k) \\ Y_3(\phi_1, \theta_1) & \cdots & Y_3(\phi_k, \theta_k) \\ Y_4(\phi_1, \theta_1) & \cdots & Y_4(\phi_k, \theta_k) \\ \vdots & \ddots & \vdots \\ Y_l(\phi_1, \theta_1) & \cdots & Y_l(\phi_k, \theta_k) \end{bmatrix}^T \begin{bmatrix} c_1 \\ c_2 \\ \vdots \\ c_l \end{bmatrix}, \tag{3.24}$$

where l is the linear SH-index defined as $l = n^2 + n + m + 1$. Equation (3.24) can be written in compact form as $\boldsymbol{p} = \boldsymbol{Y}\boldsymbol{c}$.

Spherical harmonic representation is defined as an infinite summation of spherical basis functions. However, as written in eq. (3.22), in practice, the order is truncated at a maximum value O instead. ZHANG et al. (2012) argues that spherical harmonics up to order $O = 46$ are required to correctly describe the spatial variation of an HRTF at 20 kHz. There are several sampling strategies on the sphere that allow more or less efficient conversions of the sampled spatial data into SH-domain for further calculations (ZOTTER, 2009). According to him, the most efficient method, the hyperinterpolation, requires $(O + 1)^2$ sampling points. For $O = 46$ this equals 2209 sampling points. Unfortunately, the hyperinterpolation is not an axisymmetric sampling scheme. The axisymmetric Gaussian grid requires $2(O + 1)^2$ sampling points, i.e., 4418 samples to describe the HRTFs correctly in every direction for the entire hearing range. Other axisymmetric sampling grids could reduce this number, as for example the IGLOO grid that would require only 2304 points according to ZHANG et al. (2012).[8]

---

[8]As the IGLOO grid requires more azimuth positions with less elevation positions per azimuth the reduction in measurement points will not necessarily lead to a reduction in measurement time when using the optimized MESM.

HRTFs are interpolated by first defining the spherical expansion vector c from the measured pressure values p from the minimization problem

$$\min_{c} \|p - Yc\|_2^2. \tag{3.25}$$

ZOTTER (2010) shows that for special sampling grids, e.g. the Gaussian or the hyperinterpolation grids, other solutions with more efficient numerical properties exist. A generalized solution to this minimization problem can be obtained from

$$c = Y^+ p. \tag{3.26}$$

The proceeding is completed by calculating the pressure values at the new sampling points using eq. (3.22). This operation can be conducted for p described in time or in frequency-domain. It is however more intuitive to conduct this operations in frequency-domain, as frequency-dependent order truncation can be applied (cf. POLLOW et al., 2012a).

It is important to remember that the spherical harmonics, and thus spherical holography, are defined only for a closed spherical surface. But the designed arc cannot provide measurements on the lower spherical cap (section 3.1.2). The missing points result in an ill-posed matrix of spherical harmonic basis functions Y. Regularization should then be used to obtain a stable solution that approximates the solution to the inverse problem (POLLOW et al., 2012a). RUFFINI et al. (2002) describes a regularization approach based on minimizing the *surface curvature* while matching the surface to the available data. This results in smoothly interpolated values in the lower cap region where measurements were not available. The solution to this minimization problem is given by

$$c = \left(Y^* Y + \overline{P}B + \mu PB\right)^{-1} Y^* p, \tag{3.27}$$

where $P = Y^*(YY^*)^{-1}Y$, $\overline{P} = I - P$, μ is a regularization parameter and $B = \operatorname{diag}(n(n+1))$, being n the order of the corresponding SH coefficient. A slightly altered version of this regularization scheme was also used by DURAISWAMI et al. (2004).

3.3.3 Range Extrapolation

The last step to obtain a continuous representation of the HRTFs in space is to describe its dependency on the radial distance. LENTZ (2007) proposed an interpolation scheme for near-field HRTFs. His method requires, however, the measurement of a complete set of HRTFs for several distinct radial distances in the near-field.

Using the principle of reciprocity, DURAISWAMI et al. (2004) argued that the HRTF can be "characterized as a solution of a scattering problem". According to them, a point source placed at the entrance of the ear canal will generate pressure $p(\phi, \theta, r, f)$ at a given position $\mathbf{x}$ equivalent to the pressure that would exist at the ear if the point source was placed at $\mathbf{x}$. The pressure field at position (ϕ, θ, r) and for the wave number $k = 2\pi f / c$ (where c is the speed of sound) can be represented as

$$p(\phi, \theta, r, k) = \sum_{n=0}^{\infty} \sum_{m=-n}^{n} \xi_{nm}(k) h_n(kr) Y_n^m(\phi, \theta), \qquad (3.28)$$

where ξ_{nm} is the potential expansion coefficient, h_n the spherical Hankel functions of order n and Y_n^m the already defined spherical harmonics (WILLIAMS, 1999).

The potential expansion vector $\boldsymbol{\xi}$ can be estimated from the pressure vector $\boldsymbol{p}$ by first obtaining the spherical expansion vector $\boldsymbol{c}$, as described in the previous section, and then applying the corresponding spherical Hankel function to each element of $\boldsymbol{c}$, as follows:

$$\xi_{nm}(k) = \frac{c_{nm}(kr)}{h_n(kr)}. \qquad (3.29)$$

Thus, for any given frequency f, the sound pressure field is entirely determined by the potential expansion vector $\boldsymbol{\xi}(z)$ (cf. POLLOW et al., 2012a). Because the moduli from Hankel functions behave approximately as an exponentially decaying curve, problems caused by division by zero will not occur. However, the exponential growth of the spherical Hankel functions for higher orders and small arguments kr can lead to noise amplification. This effect is commonly deal with using a frequency-dependent order truncation.

As discussed in DURAISWAMI et al. (2004), a sufficient spatial resolution is required to capture the pressure field and the required spatial resolution is proportional to the frequency. Moreover, this method will only work if all sources are contained within a spherical surface S of a small radius and the desired interpolated/extrapolated points lie outside of S. In this case, the HRTF is obtained by applying eq. (3.28) to the new desired position.

POLLOW et al. (2012a) compared the range extrapolation technique with the near-field measurements conducted by LENTZ (2007) and was able to verify that this technique produced extrapolated HRTFs that matched the measured HRTFs.

3.4 Results

This section begins with a comparison of the HRTF measurement setup presented in this chapter with a previously constructed sparse-type setup. The measurements for this first comparison were made with an artificial head. As this setup was designed for the measurement of individual HRTFs, the section then concludes showing the HRTFs and HRIRs of one of the 16 individuals that have so far been measured with the new system.

3.4.1 System Comparison

As a proof of concept the HRTFs of an artificial head were initially measured. A measurement with an artificial head offers the obvious advantage that the subject under test does not move itself during measurement and can be precisely positioned. On top of that, a comprehensive HRTF dataset of the same artificial head had already been acquired with the measurement system described by ARETZ (2012) and depictured in fig. 3.14(a). A Gaussian sampling grid of order 70 with 9800 points was used, which approximately corresponds to an angular resolution of 2°. This measurement was reported to have taken around four hours to complete using sequential exponential sweeps of 16384 samples played back at a sampling rate of 44.1 kHz.[9] The measurement was conducted at a distance of 1.75 m, different than the 1 m used with the new system. Therefore, all measured HRIRs had its phase and amplitude corrected using the Green's function to the distance of 1 m, under the assumption that the HRTFs are already in far-field (BRUNGART and RABINOWITZ, 1999).

The measurement with the new system was made using a Gaussian sampling grid of order 48, however with the lowest eight elevation points missing, which results in 3840 measurement points (cf. fig. 3.14(b)). The interleaved sweeps had a length of 59094 samples, covering the frequency range from 200 Hz to 20 kHz at a sampling rate of 44.1 kHz. The waiting time between sweeps was 35 ms. The total measurement time was just short of six minutes.

The spatial, temporal and frequency characteristic of the data obtained with both systems is compared. Figures 3.15 to 3.18 display at the top balloon plots of the measured data and at the bottom the same data

---

[9]As this system can only measure HRTFs at one hemisphere, the measurement had to be conducted in two stages, turning the head upside-down in the second stage, what is not viable for individual HRTF measurements.

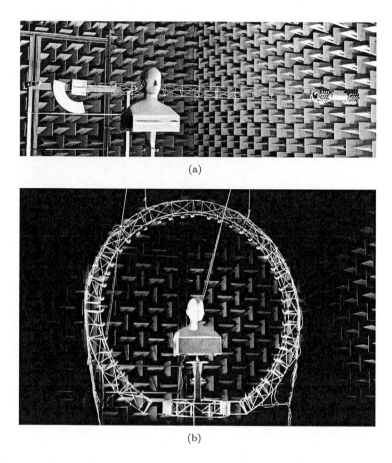

(a)

(b)

Figure 3.14: Artificial head being measured with (a) the HRTF measurement
setup previously developed at the Institute of Technical Acoustics (RWTH Aachen), composed of a single loudspeaker placed
at a rotating arm, and (b) the individual HRTF measurement
setup presented in this thesis, with a supporting arc and 40
drop-like loudspeakers.

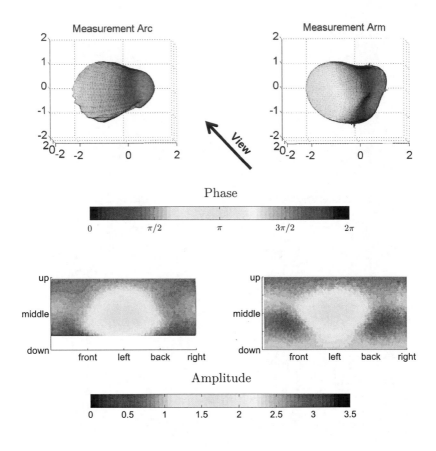

Figure 3.15: Comparison of HRTF measurement setups. The same artificial
head was measured with both the measurement arc described
in this chapter (left) and with the vintage measurement arm
described in (LENTZ, 2007) (right). The two top figures are
balloon plots, where the balloon's radius represents the amplitude of the HRTF and the color its phase. The arrow indicates
the head's view direction. The two bottom plots show the
amplitude values of both measurements plotted in an exploded
view. Plots are made for the frequency of 500 Hz.

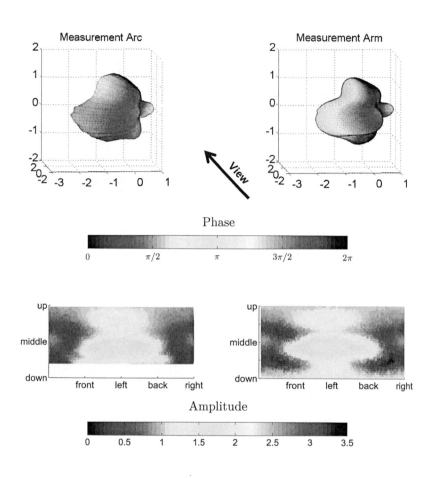

Figure 3.16: Comparison of HRTF measurement setups. The same artificial
head was measured with both the measurement arc described
in this chapter (left) and with the vintage measurement arm
described in (LENTZ, 2007) (right). The two top figures are
balloon plots, where the balloon's radius represents the amplitude of the HRTF and the color its phase. The arrow indicates
the head's view direction. The two bottom plots show the
amplitude values of both measurements plotted in an exploded
view. Plots are made for the frequency of 1000 Hz.

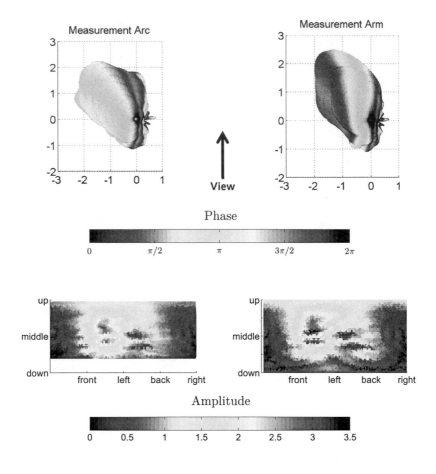

Figure 3.17: Comparison of HRTF measurement setups. The same artificial head was measured with both the measurement arc described in this chapter (left) and with the vintage measurement arm described in (LENTZ, 2007) (right). The two top figures are balloon plots, where the balloon's radius represents the amplitude of the HRTF and the color its phase. The arrow indicates the head's view direction. The two bottom plots show the amplitude values of both measurements plotted in an exploded view. Plots are made for the frequency of 4000 Hz.

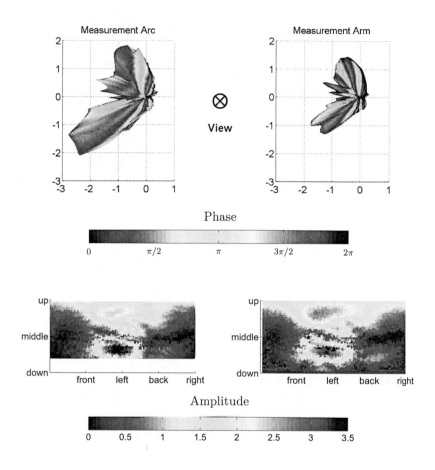

Figure 3.18: Comparison of HRTF measurement setups. The same artificial head was measured with both the measurement arc described in this chapter (left) and with the vintage measurement arm described in (Lentz, 2007) (right). The two top figures are balloon plots, where the balloon's radius represents the amplitude of the HRTF and the color its phase. The arrow indicates the head's view direction. The two bottom plots show the amplitude values of both measurements plotted in an exploded view. Plots are made for the frequency of 8000 Hz.

projected at a 2D surface. The radius of the balloon plot represents the amplitude of the HRTF and the color its phase. The arrow in between the plots shows the artificial head's view direction.

The plots at 500 Hz (fig. 3.15) make it evident the the lack of measurement data at the lower cap. To directly extract these missing values from the measured data has high uncertainty involved. For the measured region, the plots do show good similarity.

For 1 kHz, the lack of the lower cap is not as evident as for 500 Hz as now the amplitude values at this region have decreased. The edge seen in the equator of the measurement with the arm occurs because of small positioning errors as the measurement had to be done in two separate stages, each measuring one hemisphere.

The plots at 4 kHz and 8 kHz show again a good overall agreement of the data. Difference in amplitude (radius of the balloon) are observed. This are, however, compensated for when using DTFs instead of HRTFs (cf fig. 3.19). At higher frequencies it is also possible to see how the energy at the side of the contralateral ear is considerably lower than for the ipsilateral side.

The phase information encodes the distance information. As the ears are shifted in relation to the center of the head, a phase variation is observed. The different phase behavior seen at these plots is caused by the fact that, at each measurement, the artificial head was not identically positioned in relation to the systems' center. The main spatial features are, however, similar throughout the whole measured frequency range when considering only the amplitude values.

Also the frequency- and time-domain characteristics of the HRTFs measured with both systems were compared and showed good agreement. Examples for four directions are shown in fig. 3.19.

In the frequency-domain, only little deviation in the higher frequencies is clearly noticeable. In the time-domain, the main difference is observed in the pre-ringing, caused by the regularized deconvolution of the signals. These artifacts can be windowed out without compromising the quality of the HRIRs.

3.4.2 Individual Measurement

This system was designed specially for the measurement of individual HRTFs. So far, 16 listeners have been measured with he system. However,

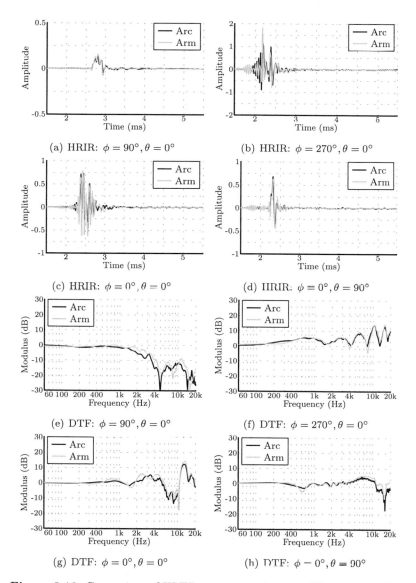

Figure 3.19: Comparison of HRTF measurement setups. The same artificial
head was measured with both the measurement arc described
in this chapter (Arc) and with the vintage measurement arm
described in (LENTZ, 2007) (Arm). HRIRs of four different
exemplary directions (a-d) and their equivalent DTFs (e-h)
show reduced variability between the measurement setups.

no localization test could be conducted to evaluate the perceptual quality of the acquired HRTFs.

The data from one of these listeners is shown below. The excitation signal was the same used for the measurement of the artificial head, described in the previous section. The spectrogram of the raw signal acquired at the left ear for one azimuthal position is shown in fig. 3.20. The 40 interleaved sweeps can be clearly seen. The last 20 sweeps have lower amplitude at higher frequencies, as could be expected once they originated from the contralateral side.

The deconvolved impulse responses can be seen in fig. 3.21. The variation in amplitude for the ipsi- and contralateral sides can be clearly seen. The small impulses present at the end of the signal are the harmonic impulse response of the first measured directions. The SNR, or better said, the peak-to-noise ratio is of approximately 80 dB when the source is directly in front of the ear and decreases to approximately 50 dB for the contralateral side. This obtained SNR is expected to be sufficient to provide high quality binaural synthesis.

The time and frequency-domain response for three positions in the median plane and one position at extreme lateral angle are shown in fig. 3.22, respectively.

3.5 Discussion

HRTF measurements were traditionally conducted in far-field, restricting the auralization to distant sources. A series of new measurements at shorter distance is required to simulate near-field effects. To avoid the need for extra measurements, the range extrapolation technique is used. It provides a spatially continuous representation of the HRTFs by using a reciprocal formulation of the modal components of an outgoing spherical wave. This results in a setup-independent and compact description of individual HRTFs, allowing the evaluation of any binaural transfer functions at any point in near- or far-field, though with some limitations due to noise and numerical instability.

The HRTF measurement setup described in this chapter was designed to meet the requirements of the reciprocal acoustic holography. This method also assumes the excitation source to be an acoustic point source. Therefore, the loudspeakers used in this setup were designed to have (approximately) an omnidirectional directivity in the entire range of

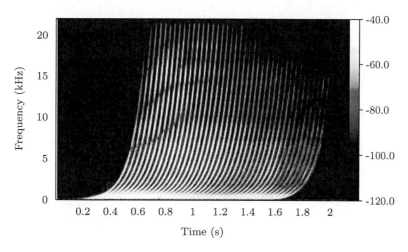

Figure 3.20: Spectrogram from the multiple exponential sweep signal acquired at the left ear of a subject for one azimuthal measurement position. The color scale is given in decibels relative to 1.

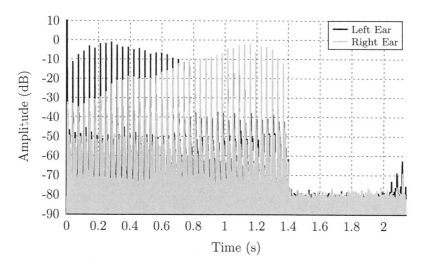

Figure 3.21: The deconvolved impulse responses obtained from the signal depictured in fig. 3.20. The peak-to-noise ratio is of approximately 80 dB for sources at the ipsilateral side and as low as 50 dB for sources at the contralateral side.

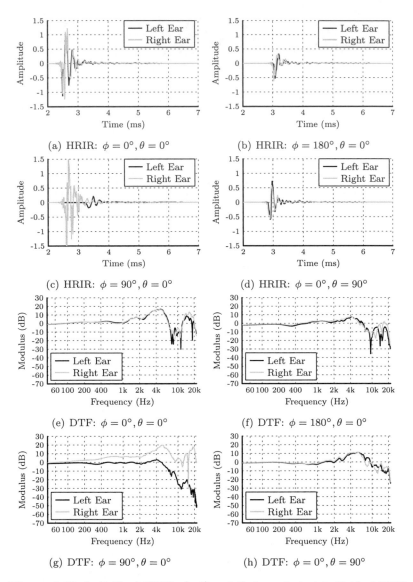

(a) HRIR: $\phi = 0°, \theta = 0°$

(b) HRIR: $\phi = 180°, \theta = 0°$

(c) HRIR: $\phi = 90°, \theta = 0°$

(d) HRIR: $\phi = 0°, \theta = 90°$

(e) DTF: $\phi = 0°, \theta = 0°$

(f) DTF: $\phi = 180°, \theta = 0°$

(g) DTF: $\phi = 90°, \theta = 0°$

(h) DTF: $\phi = 0°, \theta = 90°$

Figure 3.22: Individual HRIRs (a-d) and their equivalent individual DTF (e-h), obtained using the new HRTF measurement system presented in this chapter.

application. A small broadband loudspeaker chassis with reduced nonlinear behavior was chosen. During measurements it was verified that this chassis presented time variance which could not be compensated for and influenced the resulting HRTFs. Thus, it is recommended for new similar setups to choose loudspeakers that are not only omnidirectional, but that are also time invariant, i.e., their sensitivity does not change during the operation. Loudspeakers can, however, have a nonlinear behavior as this effect can be compensated for by performing the exponential sweep measurement.

According to acoustic holography theory, it is advisable to conduct measurements as close as possible to the scattering objects, i.e. the head and torso, and extrapolate the HRTF outwards. On that account the radius of the measurement arc was chosen to be 1 m. Results published at a later stage by POLLOW et al. (2012a) did not show any difference between inward and outward extrapolation of HRTF measurements. They were, however, able to show that the acoustic holography method is much more exact than other published methods for the HRTF near-field compensation.

Acoustic holography assumes that all scattering objects are contained inside a spherical surface S of small radius and that all sound sources are located outside S. For this reason, the supporting arc was constructed using a thin metal rod truss structure, which is supposed to be acoustically transparent, and the form of the loudspeakers was chosen to minimize reflections. During measurements, though, it was observed that unwanted reflections from neighboring loudspeakers and structure-borne sound from the arc were still present. Therefore, the design of the arc and the loudspeakers should be reviewed. The loudspeakers should be integrated into a rigid arc. Moreover, it was also verified that the mechanical stability of the chosen trellis structure was not sufficient.

The new setup was, however, not only used for adequate range extrapolation measurements. The other purpose of this new setup was to measure a sufficient number of points on the sphere to adequately describe the HRTF in the *shortest time* possible. This objective was achieved by optimizing the excitation signal used for the measurement. MAJDAK et al. (2007) originally proposed to overlap the exponential sweep signals in order to reduce measurement time without compromising the obtained SNR. This thesis extends their multiple exponential sweep method (MESM) in two ways: 1) by relaxing the overlapping requirements and 2) by making better use of the HRIR's temporal structure. The original requirement for signal overlapping was that all harmonic IRs should decay below background noise level before the previous desired IR occurs.

This requirement was relaxed so that no harmonic IRs shall fall into a predefined avoid zone containing the desired IR. Observing the temporal structure of an HRIR measured in a typical (hemi-)anechoic chamber, one notes that only the initial portion of the IR describes the HRIR, being the rest of the signal composed by reflections due to objects in the room or to room boundaries. As these unwanted reflections must be windowed out, these regions can be used as place-holders for harmonic IRs of other directions. These reflections are responsible for a longer reverberation time of the chamber, which in turn limits the waiting time between subsequent sweeps. The size of the hemi-anechoic chamber used with this setup was *reduced* by building a wall of absorbing material near the arc, thus reducing the size of the chamber's IR and consequently speeding up the measurement. Altogether, these optimization allowed the HRTF measurement in 3840 directions in less than six minutes,[10] which would take approximately 12 min with the original MESM and over 1.5 h using the sequential method.

The post-processing of the raw IR is executed in accordance with the latest research found in the literature, as e.g. in HAMMERSHØI and MØLLER (2005). One should only keep in mind that the HRTFs were no longer measured under far-field assumption and therefore do not allow the distance effect to be compensated for simply by the Green's function, as was the case with the free-field HRTF. To serve as input data for spherical holography, the HRTFs should be multiplied by the transfer function of a point source placed at the acoustic center of the corresponding loudspeaker to a point receiver placed at the origin of the head coordinate system.

Results of measurements showed the presence of a spatial ripple in the resulting HRTFs. The effect of these ripples when calculating the spherical wave spectrum are negligible in amplitude, but have a strong effect in the phase, as described in (KRECHEL, 2012). These ripples are caused either by reflection artifacts still present in the measurement or by the fact that the subject's rotation axis, when placed over the turntable, cannot be precisely aligned with the arc's symmetry axis. In the first case, ripples are eliminated by adequate time windowing. In the second case, the proposed way to mitigate this effect is to place the reference microphone at the center of the turntable and conduct a regular measurement rotating it, just like the subjects are, and using the transfer function obtained for each direction to equalize the HRTF of the correspondent azimuthal direction.

---

[10]Being the fastest HRTF measurement setup known to the author.

To find the spherical expansion vector required for interpolation, it is extremely important to know the position of the loudspeakers as precisely as possible. Therefore, a calibration system was developed using only the turntable and two microphones placed at a known distance from each other. This system gives not only the exact position of the loudspeakers, but also the orientation of the bar holding the microphones, the distance to the rotation axis, the latency of the sound card and the speed of sound at the time of measurement (KRECHEL, 2012).

A method to accomplish interpolation and range extrapolation of the HRTF based on acoustic spherical holography was described. This method assumes the HRTF has a continuous and smooth distribution in space, described by a finite number of spherical harmonics. Ideally, the number of available sampling points should be sufficient to unambiguously represent all needed spherical harmonic basis functions. This restriction is hardly achievable in practice and a small amount of spatial aliasing is to be expected. ZHANG et al. (2012) argues that spherical harmonics up to order $O = 46$ are required to correctly describe the spatial variation of an HRTF at $20\,\mathrm{kHz}$. POLLOW et al. (2012c) show that displacing the origin of the spherical coordinate system to the acoustic center of reciprocal HRTF pressure field allows a more compact description of the HRTF with fewer spherical coefficients (therefore, with a lower maximum order). Consequently, fewer measurement points are also required. A potential vector $\boldsymbol{\xi}$ for the head coordinate system can be obtained from the potential vector $\boldsymbol{\xi}'$, with origin placed at the entrance of the ear canal, by the translation operation in the spherical coordinates, thoroughly described in (ZOTTER, 2009, pp. 36-50).

To avoid the effects of spatial aliasing—and also improve the overall measurement quality—efforts have been made to develop a method that can measure the HRTF on a continuous surface or at least along a circle (AJDLER et al., 2007; FUKUDOME et al., 2007; ENZNER, 2009). These techniques are, however, not as robust to nonlinearity as the correlation measurement technique using the exponential sweep described in section 3.2. KRECHEL (2012) described an approach to dynamically acquire the HRTF along a circle using the correlation technique, allowing a further considerable speedup in comparison to the sequential measurement technique. In this method, the listener is continuously turned while the excitation signals are being played. Even though this continuous movement breaks the assumption of time invariance implicit to the correlation technique, it is plausible to assume that the system is "almost" invariant at each small time interval, while one frequency is been played. A post-processing step is then necessary to compensate for this continuous movement.

The recently developed *Compressive Sampling* theory was also studied as a way to reduce the number of required sampling points and to avoid spatial aliasing (MASIERO and POLLOW, 2010). Compressive sampling proposes a new framework on how to effectively sample information with a reduced number of sensors. The main idea behind this concept is that if the information to be sampled can be sparsely described in a space that is incoherent to the measurement space, then this information can be restored by ℓ_1 minimization. Unfortunately, compressive sampling could not be applied to the HRTFs using the spherical harmonic basis functions, as especially high frequency HRTFs cannot be considered sparse in the SH-domain. A set of basis functions extracted from the "principal component analysis" of a group of individual HRTFs, similar to the basis described in (KISTLER and WIGHTMAN, 1992), but spanning the whole sphere, might be a good candidate for an incoherent representation domain. Other possible basis would be spherical wavelets (FREEDEN and WINDHEUSER, 1997) or the Slepian functions (SLEPIAN, 1964).

4

Binaural Reproduction using Headphones

The reproduction of binaural signals via headphones is straightforward as headphones are able to deliver each binaural channel independently to each ear. The headphone reproduction does add spectral coloration to the reproduced sound, but at first glance it seems as if this effect can easily be mitigated by an equalization filter. There are, however, some difficulties involved in the design of such a filter, which is obtained from the inverse of the *headphone transfer function* (HpTF). First, the HpTF varies for each listener. Therefore headphone equalization filters must be shaped individually. Second, at high frequencies the HpTF is strongly dependent on the headphone fitting and therefore the equalization filter should be robust to (small) fitting variations. Third, HpTF are usually not minimum-phase, i.e. they contain all-pass components that when inverted result in a noncausal equalization filter.

The HpTF and HRTF are commonly measured with a microphone placed at the entrance of the ear canal. However, a correct binaural reproduction occurs when the sound pressure at the listener's eardrums is ideally matched. MØLLER (1992) proposed a measurement technique to verify whether a given set of headphones is able to provide an authentic binaural reproduction. This technique is applied in this chapter to verify the adequacy of the used headphones.

Measured HpTFs were evaluated with regard to the inter-subject variability and intra-subject variability to the headphone fitting. For frequencies up to 4 kHz a low variability was observed in both cases. Above this frequency , standing waves start to build up inside the cavity and thus the resulting pressure at the listeners' eardrums becomes strongly dependent on the geometry of the listener's ear and on the headphone fitting (SCHMIDT, 2009). This high variability for subjects

---

†Part of the results presented in this chapter have been previously published in
- MASIERO and FELS (2011b);
- MASIERO and FELS (2011a);
- FELS and MASIERO (2011).

corroborate for an individual equalization. To reduce the variability between fittings, the listeners should fit the headphones themselves at the most comfortable position.

This chapter starts by characterizing adequate headphones and microphones for an authentic binaural reproduction. Then the influence of headphone fitting on the measured individual HpTF is analyzed. Furthermore, an individual headphone equalization technique, perceptually robust to small variations in headphone fitting, is presented. The chapter concludes with a discussion of the obtained results.

4.1 Headphone Type

According to MØLLER (1992), ideal binaural reproduction via headphones is obtained if the headphone listening condition is equal to the free-field listening condition. He shows that for this to be true, the acoustical impedance seen from the ear canal should be the same for the two conditions. To verify if a headphone fulfills this requirement, he defined the *pressure division ration* (PDR) as

$$\text{PDR}(z) \equiv \frac{H_{\text{FF}}^{\text{ED}}(z)/H_{\text{FF}}^{\text{EC}}(z)}{H_{\text{HP}}^{\text{ED}}(z)/H_{\text{HP}}^{\text{EC}}(z)} \tag{4.1}$$

where $H_{\text{FF}}^{\text{ED}}(z)$ is the transfer function from a free-field sound source (FF) to the listener's eardrums (ED), $H_{\text{FF}}^{\text{EC}}(z)$ is the transfer function from a free-field sound source to the microphones at the entrance of the ear canal (EC), $H_{\text{HP}}^{\text{ED}}(z)$ is the transfer function from the headphones (HP) to the listener's eardrums, and $H_{\text{HP}}^{\text{EC}}(z)$ is the transfer function from the headphones to the microphones at the entrance of the listener's ear canal (cf. fig. 4.1).

The idea behind the PDR is to verify if equalized headphones, when playing a binaural signal, can generate the same sound pressure at the listener's eardrum that would be generated by the original free-field sound source. This will only occur if $\text{PDR}(z) = 1$. MØLLER et al. (1995b) showed that all headphones tested by them fulfilled this criterion for frequencies below 2 kHz and that only a few fulfilled this criterion for frequencies between 2 and 7 kHz. They named the headphones belonging to the second group as *free-air equivalent coupling* (FEC) headphones as they "do not disturb the radiation impedance as seen from the ear" (MØLLER, 1992). The methodology used by them did not allow a reliable measurement of the PDR for frequencies above 7 kHz. Later investigations confirmed that FEC headphones cause the smallest variation on the impedance seen from the ear canal when compared to the free-field listening condition (KLEBER and VORLÄNDER, 2001; CRUZADO, 2002).

Two candidate headphones—one having an electrodynamic transducer (Sennheiser HD-600) and the other having an electrostatic transducer (Stax SRλ), both of open-type—were tested regarding their PDR by measuring all four TFs defined in eq. (4.1) using an artificial head equipped with an IEC 711 ear simulator, i.e. with an artificial ear canal. For frequencies below 10 kHz both measured headphones complied with the FEC criterion, agreeing with the results presented by VÖLK (2011b). The electrodynamic headphones were chosen for further tests.

The PDR may also vary due to the headphone fitting and the type of microphones that are used. OBEREM (2012) investigated both aspects. She conducted repeated measurements with the same setup described above, replacing the headphones and microphones at each new measurement. The results show that the miniature microphone placed with an ear plug at the entrance of the ear canal (also called acoustic *meatus*) yields the best results, as can be seen in fig. 4.2. Therefore, further measurements were conducted using miniature microphones placed at the entrance of the blocked ear canal.

4.2 Variability of Headphone Fitting

The variability of headphone transfer function (HpTF)[1] due to the fitting has been extensively investigated (TOOLE, 1984; MØLLER, 1992; CRUZADO, 2002; VÖLK, 2011a). Furthermore, PAQUIER and KOEHL (2010) confirmed the importance of headphone variations as they were able to verify that these spectral differences are audible.

A series of individual HpTF measurements was carried out with 15 listeners. Generally speaking, it can be observed that interindividual HpTFs have a low variability up to approximately 4 kHz. In this frequency region, where the headphone works as an acoustic cavity, just a constant level variation is observed, caused by variable leakage, as described by TOOLE (1984). For higher frequencies, resonances are observed which vary with headphone fitting and the geometry of the listeners' ears. This higher variation occurs for two reasons:

---

[1] All HpTF plots have the y-axis displayed in dB relative to 1 Pa/V and only results of the left ear are displayed.

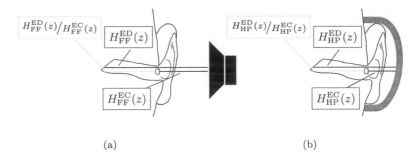

(a)   (b)

Figure 4.1: Sketch of transfer paths required for the PDR calculation. (a)
The free-air condition, measured with a loudspeaker in free-field
and (b) the headphone condition.

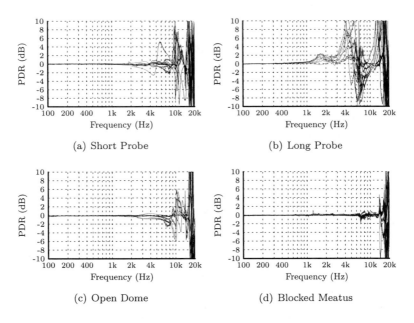

Figure 4.2: PDR measured with an electrodynamic open headphone
(Sennheiser HD-600) and with (a) a short probe microphone,
(b) a long probe microphone, (c) a miniature microphone in
an open dome, and (d) a miniature microphone in an ear plug
blocking the meatus.

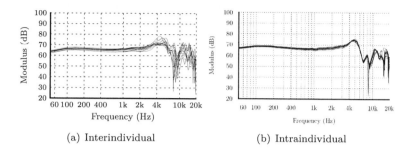

(a) Interindividual   (b) Intraindividual

Figure 4.3: HpTF measured at the left ear with the Sennheiser HD-600 headphone for (a) fifteen different subjects and (b) one single subject with fifteen repetitions (at each new measurement the subject replaced the headphone for a comfortable fit). The good agreement at the intraindividual measurement was also observed with other subjects.

1. Because at this frequency range standing waves start to build up inside the headphone's cavity.

2. Because of the size of the external ear structures (SCHMIDT, 2009, p. 84)—meaning that in this region the HpTF behavior is highly individual, as can be seen in fig. 4.3(a).

Each individual repeated the HpTF measurement 15 times. For every new measurement the listener was instructed to place the headphone at its most comfortable fit. The result from the interindividual measurement for one exemplary listener can be seen in fig. 4.3(b) and confirmed that low measurement variability can be achieved if listeners are allowed to fit the headphone themselves at a comfortable fit. MASIERO and FELS (2011b) showed that if listeners are instructed to place the headphones at extreme positions and not only at comfortable positions, the variability of the measured HpTF increases especially in the frequency range above 4 kHz.

4.3 Robust Individual Equalization

The variability results described in the last section agree with the results presented by HAMMERSHØI and MØLLER (2005), who claim that "it can be seen that the variations between measurements are much less than variations between subjects. The low variation in the repeated measurements means that an individual headphone (equalization) filter can be reliably designed. The high variation in transfer functions across subjects means that it probably should be designed individually". Therefore, to achieve a robust equalization, the equalization filter is constructed from the average of several individual HpTF measurements,[2] always completely removing the headphones in between measurements.

Headphone repositioning will cause variations that itself affect the equalized frequency response and appear as peaks or dips in the higher frequency range. BUCKLEIN (1981) conducted speech intelligibility tests and showed that human listeners are more sensitive to spectrum irregularities in form of peaks than to equivalent valleys. Assuming that this behavior extends also to spatial perception, headphone equalization filters should also avoid the occurrence of resonance peaks in the equalized response. This can be achieved by applying a notch smoothing algorithm to the amplitude spectrum (cf. MÜLLER, 1999, p. 192), which first smooths the entire frequency response and then compares it with the original function. At regions where this difference is higher than a given threshold, a cross-fading is made, thus locally smoothing the original function. Optionally, a softer smoothing can be done throughout the whole frequency spectrum afterwards.

Regarding the filters' overall gain, ideally, the equalization filter should not alter the loudness of the reproduced signals; but loudness measurements are dependent on the type of signal being used. For broadband signals, if the overall sound pressure level is kept constant, negligible variation on the loudness values should be observed. Therefore, the smoothed average HpTF is normalized by the root mean square value of the frequencies showing low variability, i.e., below 4 kHz. The applied weight should be the average of the RMS from both ears to allow the proper equalization of the interaural level difference.

As with any other equalization filter, care must be taken at frequencies outside the roll-off frequencies as correction at these regions may lead to very large gains that can produce undesired nonlinearity in the equalized response. Likewise, to equalize a headphone at low frequencies (below approximately 100 Hz) a very long FIR equalization filter is required. Since these equalization filters are aimed for use in real time virtual reality systems, it is of interest that these filters are kept short in order to avoid extra latency. As the low frequency range does

---

[2]Spectral average should be obtained independently for the amplitude spectrum and the group delay to avoid the unwanted phase canceling effect.

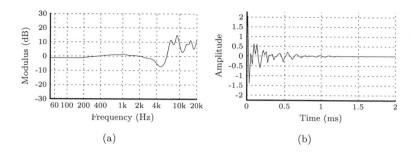

Figure 4.4: Individual headphone equalization filter calculated from the average of seven HpTFs. A notch smoothing algorithm was applied followed by a 1/6 octave smoothing (a). The time response (b) is obtained from the minimum-phase spectrum of (a).

not contribute to localization and the HpTF variation due to individual fitting in this frequency range is very low, this frequency region can be left untouched. This is done by substituting the frequencies below the first observed maximum of the HpTF with a constant line of the same value as the amplitude of the first maximum.

The last step is to invert the HpTF. MINNAAR et al. (1999) discuss that HpTFs generally contain all-pass components that, when inverted, will drive the equalization filter to be noncausal. This effect could be compensated for by inserting a delay in the equalization filter or by equalizing only the minimum-phase component of the HpTFs. With the headphone fitting variation, the first option might lead to the compensation of a nonexistent all-pass section while the second option will not correct the present all-pass sections. MINNAAR et al. (1999) suggest "that it will be more safe not to equalize for an all-pass that is there than to equalize for an all-pass that is not there." Therewith, only the magnitude spectrum of the smoothed average HpTF was inverted and the equivalent minimum-phase spectrum was obtained using the Hilbert transform, thus producing a causal and compact equalization filter.

4.4 Results

Figure 4.4 shows an example of an individualized headphone equalization filter. This filter was calculated from the average of seven individually measured HpTF magnitude spectra, each with a new headphone fit-

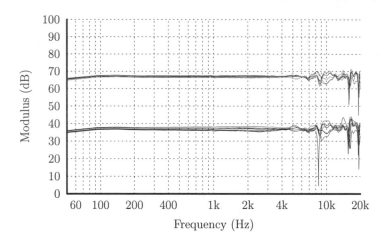

Figure 4.5: Equalization response for an individual headphone equalization
filter. Seven HpTFs were averaged to generate the filter. The
upper curves are obtained by multiplying the equalization filter
with the same HpTFs used for the calculation. The lower curves
are obtained by multiplying the equalization filter with the seven
other HpTFs measured for the same listener. These curves are
shifted by -30 dB for clarity.

ting. A notch smoothing algorithm was applied followed by a 1/6 octave logarithmic smoothing. The low frequency correction is truncated
at approximately 200 Hz and the time response is obtained from the
minimum-phase spectrum.

Applying this filter to the original HpTFs will result in a considerably
flat equalized headphone response, except for spectrum dips at high
frequencies, as depicted in the upper curves of fig. 4.5. This is, however,
not a realistic situation as these HpTFs are the same ones used for the
filter calculation. A realistic situation can be observed if the equalization
filter is applied to a different set of HpTFs from the same listener. In
this case, the variability increases slightly, as can be seen in the lower
curves of fig. 4.5.

4.5 Discussion

Headphone Transfer Functions (HpTFs) were measured using an artificial head and individual subjects, confirming that for low and middle

frequencies only small level variations are present while for the high frequencies very individual resonance patterns are found. Low measurement variability is achieved if subjects are allowed to fit the headphone themselves at the most comfortable position. This underlines the fact that individual equalization should be used when possible.

A robust equalization filter design is proposed, inverting the average of the magnitude spectra of several individually measured HpTFs. A notch smoothing algorithm is applied to avoid high peaks at the spectrum of the equalization filter. The peaks in the HpTF—dips in the equalization filter—are not altered as the human hearing system is more sensitive to irregularities in form of peaks than to irregularities in form of dips. To avoid a noncausal equalization filter, only the minimum-phase component of the averaged HpTF is inverted. Furthermore, the headphone is not equalized for very low frequencies to keep the FIR equalization filter short.

The effectiveness of this filter calculation strategy is evaluated in chapter 6. It was verified that adequate individual equalization provide realistic binaural reproduction, in the sense that the listeners cannot differentiate between the real binaural signal generated by a loudspeaker and the virtual binaural signal synthesized with the individual HRTF and played back via individually equalized headphones.

If, however, an individual equalization cannot be carried out, LARCHER et al. (1998) suggest the use of diffuse-field-compensated headphones with no additional equalization and diffuse-field equalized binaural signals. These signals can be synthesized by using DTFs instead of HRTFs.

5

Binaural Reproduction using Loudspeakers

If loudspeakers are used to reproduce a binaural signal, left and right signal will arrive mixed together at the listener's left and right ear, thus destroying the binaural cues and the spatial impression. To reestablish these cues *crosstalk cancellation* (CTC) filters, presented in section 5.1, are used to generate (from the input binaural signal) transaural signals to be fed to the loudspeakers which should interact to reproduce the binaural signal at the listener's ears with sufficient channel separation (defined in section 5.2).

The crosstalk cancellation is achieved by means of constructive and destructive wave interference. At some frequencies CTC filters may display elevated gains that might require the loudspeakers to reproduce very high sound pressures; only to be later (partially) canceled at the listener's ears (TAKEUCHI and NELSON, 2007). To avoid clipping and distortion at these frequencies, the overall gain of the CTC filter has to be reduced, causing the dynamic range of the reproduced binaural signal to shrink. According to NELSON and ROSE (2006), these frequencies with extreme high energy will result in a poorly damped ringing behavior in the time-domain. This behavior will also occur in the spatial-domain, leading to a very narrow region with adequate binaural reproduction—the so called *sweet spot*.

The sweet spot can be enlarged by using a loudspeaker array with the high-frequency (tweeter) sources placed close to each other and the low-frequency sources (woofer) placed opposite to each other (BAUCK and COOPER, 1996; TAKEUCHI and NELSON, 2007). However, to apply this loudspeaker placement strategy to an immersive virtual reality (VR) system, where the listener is constantly moving and consequently the sweet spot is also constantly shifting, would not be viable as a very large number of high frequency transducers distributed all around the

---

†The results presented in this chapter are an extension of the results published in

- MASIERO; FELS, and VORLÄNDER (2011b);
- MASIERO and VORLÄNDER (2012).

reproduction space would be required. Therefore, all loudspeakers in this chapter are assumed to be broad-band sources.

CTC filters can be realized in various ways: either analog or digital, FIR or IIR, with or without room equalization. This thesis focuses on CTC filters for immersive virtual reality applications, which must be constantly updated to compensate for the listener's motion, should be as short as possible to save calculation power, and should introduce the shortest possible latency in the system to allow fast reaction to listener movements. For such applications, the use of digital FIR filters is preferred as it allows efficient filter updates with fast filter calculation (LENTZ, 2007).

Digital CTC filters can be calculated either in time or in frequency-domain. Calculations in time-domain produce strictly causal filters. On the other hand, frequency-domain calculations are computationally more efficient, but may result in noncausal filters (see discussion in section 5.3). Apart from that, there is no substantial difference between the results achieved using the two methods (PARODI, 2008). In section 5.3 a general framework is introduced for the calculation of digital CTC filters with causality constraints in the frequency-domain.

Causal filters could also be achieved using a minimum-phase version of the HRTFs, as did by GARDNER (1997). PARODI (2008) compared the performance of this generic CTC filter proposed by GARDNER (1997) with time and frequency-domain least-mean-square approximations and concluded that the generic CTC filters provided reduced channel separation performance. The generic CTC filter calculation proposed by GARDNER (1997) can also only be applied to two loudspeaker CTC filter. The framework presented in this chapter can be applied to generate CTC filters for an unlimited number of loudspeakers.

KIRKEBY et al. (1998b) showed that CTC filters have infinitely long impulse responses (IRs) even when they are derived from a set of finite head-related transfer functions (HRTFs). They proposed the use of Tikhonov regularization to control undesirably large peaks in the frequency response of the CTC filters. As these peaks are responsible for weakly damped ringing behavior in time-domain, the use of regularization also reduces the length of the CTC filters. The side-effect of regularization is the appearance of unwanted noncausal artifacts in both the CTC filters and the resulting ear signals. Taking into account the results of minimum-phase regularization, the presented framework is extended to force the filters resulting from the regularized inverse problem to be causal.

To avoid filter instability as users rotate their head inside an immersive VR environment, LENTZ (2006) suggested the use of four loudspeakers from which only two loudspeakers are active at a time, depending on the orientation of the listener's head. In section 5.4 an improved solution for the filter switching strategy is presented which integrates spatial fading in the filter design stage and makes it possible to smoothly switch between active loudspeakers.

The CTC filter calculation framework presented in this chapter is based on the knowledge of the transfer-path between loudspeakers and listener's ears, i.e., the HRTF. However, no restriction is made if these HRTFs have to be individually measured. The influence of individualized HRTFs in localization performance with CTC systems will be later discussed in section 6.2.

In this chapter some results will be discussed in regard to the obtained channel separation, defined in section 5.2, which is commonly referred to in literature as a quality predictor for CTC filters. In section 6.2 the relationship between channel separation and localization performance of a CTC system is investigated.

5.1 CTC Reproduction System

Figure 5.1 shows the setup of a CTC system with two loudspeakers. The transmission path from the loudspeakers to the listener's left and right eardrums can be written in the frequency-domain as

$$E_L(z) = H_{1L}(z)V_1(z) + H_{2L}(z)V_2(z), \tag{5.1a}$$
$$E_R(z) = H_{1R}(z)V_1(z) + H_{2R}(z)V_2(z), \tag{5.1b}$$

where $E_L(z)$ and $E_R(z)$ are the signals at the listener's ears, $V_1(z)$ and $V_2(z)$ are the signals fed to the loudspeakers and $H_{nL}(z)$ and $H_{nR}(z)$ represent the acoustic path from the nth loudspeaker to the left and right ears, respectively.

Equation eq. (5.1), can be written in matrix formulation as

$$\begin{bmatrix} E_L(z) \\ E_R(z) \end{bmatrix} = \begin{bmatrix} H_{1L}(z) & H_{2L}(z) \\ H_{1R}(z) & H_{2R}(z) \end{bmatrix} \cdot \begin{bmatrix} V_1(z) \\ V_2(z) \end{bmatrix} \tag{5.2}$$

or

$$e = Hv, \tag{5.3}$$

where elements of e are the signals at the listener's ears, elements of H (called *acoustic transfer matrix*) describe the acoustic propagation

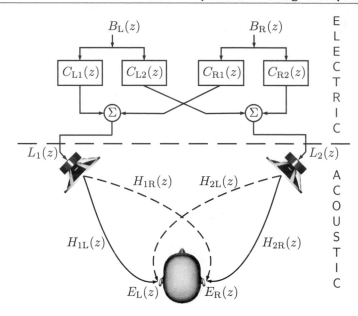

Figure 5.1: Diagram of a binaural reproduction system using loudspeakers, i.e., a crosstalk cancellation (CTC) system. The CTC filters are shown in the upper part and the acoustic paths are shown in the lower part of the figure. The solid and dashed lines show the direct and the crosstalk paths, respectively.

paths, and the elements of v are the loudspeaker signals, all in frequency-domain.[1]

The crosstalk paths can be canceled out using an adequate filter structure. This should be always placed between the input binaural signal and the loudspeakers (see fig. 5.1), and can be represented as matrix C, the so-called *crosstalk cancellation matrix*, such that

$$v = Cb, \qquad (5.4)$$

where the elements of b are the left and right binaural signals to be presented, resulting in the complete transmission path

$$e = HCb. \qquad (5.5)$$

---

[1] All equations presented in this section are in frequency-domain. They can be recast, however, in a time-domain representation, as discussed in section 5.3.

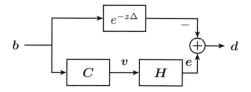

Figure 5.2: The crosstalk cancellation problem displayed as a block diagram.

A correct binaural reproduction is achieved when, apart from a time delay, the binaural signal b is exactly reproduced at the listener's ears. As discussed in section 2.2, this problem can be studied as a minimization problem, viz. the minimization of the reproduction error

$$d = \left(e - b \cdot e^{-z\Delta}\right), \qquad (5.6)$$

where Δ is a time delay proportional to the acoustic lag between loudspeakers and listener position. The block diagram form in fig. 5.2 shows the minimization problem that occurs while obtaining optimal CTC filters.

Substituting eq. (5.5) in eq. (5.6) one has

$$d = \left(HC - I \cdot e^{-z\Delta}\right) b, \qquad (5.7)$$

which highlights the dependence of the optimal CTC filter on the input binaural signal. KIRKEBY and NELSON (1999) suggest substituting b by a delta function, thus obtaining a filter for the "worst case" scenario where the input binaural signal contains energy in the entire frequency spectrum.

For the two loudspeaker setup shown in fig. 5.1, the crosstalk cancellation matrix that minimizes the reproduction error can be obtained from eq. (2.7). Assuming that H is invertible, this is given by

$$C = H^{-1}e^{-z\Delta}. \qquad (5.8)$$

The binaural signals do not necessarily have to be reproduced using only two loudspeakers (BAUCK and COOPER, 1992). If N loudspeakers are used instead, H expands to

$$H = \begin{bmatrix} H_{1L}(z) & H_{2L}(z) & \cdots & H_{NL}(z) \\ H_{1R}(z) & H_{2R}(z) & \cdots & H_{NR}(z) \end{bmatrix}.$$

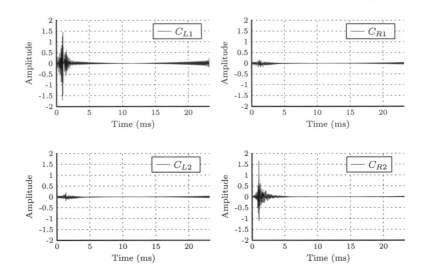

Figure 5.3: Time response of C for two loudspeakers placed at $\phi = \pm 45°$ calculated with the regularized equation eq. (5.24) using $\mu = 0.005$ for all frequencies and $\Delta = 3.4\,\text{ms}$. Noncausal oscillations are clearly visible in all four filters, even though a time delay proportional to the distance between loudspeakers and head was used.

H now represents an underdetermined system and the CTC filters obtained by a least-squares minimization are given by (see appendix B)

$$C = H^* \left(H H^* \right)^{-1} e^{-z\Delta}. \tag{5.9}$$

The CTC formulation could easily be expanded for multiple listeners by concatenating e, H, and b without actually altering the filter calculation scheme (KIM et al., 2006; MASIERO and QIU, 2009).BAUCK and COOPER (1992) studied a number of multiple listeners CTC configurations, for instance a setup with fewer loudspeakers than listener's ears (overdetermined system) and a very interesting setup for cinema application that uses one central loudspeaker and several distributed dipole loudspeaker placed behind the listeners' heads. This chapter, however, will focus only on the one listener setup.

5.2 Channel Separation

The channel separation (CS) has been proposed to describe the quality of CTC systems (GARDNER, 1997) and has been defined as the logarithmic difference between the signals at the ipsilateral and the contralateral ear (BAI and LEE, 2006). Thus, assuming

$$b = \begin{bmatrix} 1 \\ 0 \end{bmatrix},$$

an ideal CTC system will generate a signal only at the left ear. Equation (5.5) then reduces to

$$e_L = H_{1L}C_{L1} + H_{2L}C_{L2}, \tag{5.10a}$$

$$e_R = H_{1R}C_{L1} + H_{2R}C_{L2}. \tag{5.10b}$$

and the channel separation for the left ear can be calculated as

$$\mathrm{CS_L} = 20\log_{10}\left(\frac{|H_{1L}C_{L1} + H_{2L}C_{L2}|}{|H_{1R}C_{L1} + H_{2R}C_{L2}|}\right). \tag{5.11}$$

The same applies to the right ear. This definition for CS suggests that a larger CS results in a better CTC system, which follows the definition used by AKEROYD et al. (2007) and QIU et al. (2009), but is contrary to the definition used by BAI and LEE (2006) and PARODI and RUBAK (2010). Note that the CS is given separately for each frequency and that the CS is usually averaged over a defined frequency range in order to obtain a single valued quality metric for the CTC system (BAI and LEE, 2006; AKEROYD et al., 2007).

Without the use of CTC filters, i.e., assuming that $C = I$, the channel separation for the left ear would be

$$\widehat{\mathrm{CS}}_L = 20\log_{10}\left(\frac{|H_{1L}|}{|H_{1R}|}\right). \tag{5.12}$$

Again, the same definition applies to the right ear. $\widehat{\mathrm{CS}}$ represents the natural CS caused by head shadowing and it is equivalent to the CS observed using a simple stereophonic reproduction system. The $\widehat{\mathrm{CS}}$ depends directly on the system's loudspeaker position. It is also frequency-dependent and it has its maximum of approximately 30 dB at higher frequencies (BLAUERT, 1997).

For an ideal CTC system, the obtained CS is expected to be substantially larger than $\widehat{\mathrm{CS}}$. However, in practical applications, the *setup* HRTFs used to design the CTC filters in C are not always identical to the *playback* HRTFs contained in H. In these cases, a poorer performance of the CTC system can be expected, as shown in section 6.2.2.

5.3 Filter Design

As discussed in the previous section, CTC filters can be calculated either in the time or in the frequency-domain. The frequency-domain solution given in eq. (5.8) can be recast in the time-domain as

$$\widehat{C} = \widehat{H}^{-1} I(\Delta). \tag{5.13}$$

where $\widehat{H}$ is the concatenation of the convolution matrices of each HRIR in H, $\widehat{C}$ is the concatenation of the impulse response from the CTC filters, and $I(\Delta)$ is a block diagonal matrix with two delayed delta functions in its diagonal (KIRKEBY and NELSON, 1999; PARODI, 2010).

If the available HRIRs are N samples long and the desired CTC filters are M samples long, then $\widehat{H}$ will be a $2(M + N - 1) \times 2M$ matrix for the two loudspeaker CTC configuration. This matrix is overdetermined and applying eq. (2.7) would result in the inversion of a single $2M \times 2M$ real matrix. The same problem occurs in the frequency-domain. Assuming that the filter length for both H and C is $M + N - 1$, would require $M + N - 1$ times the inversion of a 2×2 complex matrix.[2]

As a matrix inversion has a computational complexity $O(n^3)$,[3] an inversion in the frequency-domain has the advantage that its computational requirements are considerably smaller than computation in the time-domain, even considering the required FFTs. This is a major advantage for real-time VR systems since these types of systems require constant filter updating. Already for medium-size filters (around 500 coefficients), it is usually more efficient to repeat the inversion of a small matrix several times, as in the case of the inversion in frequency-domain, than to invert a large matrix only once, which would be done when obtaining the filters directly in time-domain.

---

[2] Please note that H and C are a three-dimensional tensor, while e, v, and R are two-dimensional tensors. As the addition and multiplication operations can be conducted independently in the frequency dimension, for each frequency, the three-dimensional tensors can be considered matrices and the two-dimensional tensors can be considered vectors.

[3] There are faster algorithms for matrix inversion with a computation complexity as low as $O(n^{2.3727})$. However, these algorithms usually only produce a considerable speedup for very large matrices.

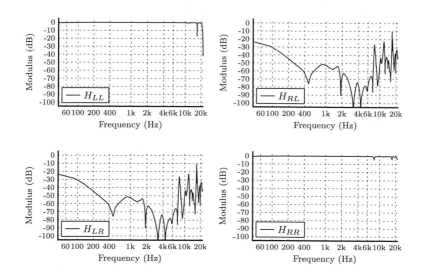

Figure 5.4: Frequency response of the complete transfer-path between the binaural signals and the ear signals for the filters shown in fig. 5.3. The diagonal elements are ideally 0 dB, the off-diagonal elements are ideally $-\infty$ dB. Deviation to the ideal result is caused by the regularization applied at the CTC filter calculation.

When calculating the CTC filter for a two loudspeaker setup in the frequency-domain, there is only one set of filters that can force the reproduction error to be exactly zero. If more than two loudspeakers are available, the transmission matrix H turns into an underdetermined matrix and in this case, there is an infinite number of CTC filters that can deliver an ideal reproduction. The least-square minimization (LSM) used so far chooses (from this infinite group of filters) the CTC filter combination with minimum Euclidean norm (energy) solution by solving the minimization problem

$$\underset{c_j}{\text{minimize}} \quad \|c_j\|_2$$
$$\text{subject to} \quad HC = I, \tag{5.14}$$

where c_j is the jth column of C. This minimization can be solved by applying the Lagrangian multipliers, as explained in appendix B.

Instead of minimizing the ℓ_2-norm, one could minimize the ℓ_1-norm of c_j, obtaining a set of filters with its coefficients sparsely distributed in the loudspeaker dimension, i.e., at each frequency the least possible number of loudspeakers will be active. This results in CTC filters whose energy is compactly distributed in the spectrum.[4] The result obtained using the ℓ_1-norm minimization resembles the optimal source distribution (OSD) setup, as TAKEUCHI and NELSON (2007) suggests using only two distinctly positioned loudspeakers for each frequency band as well.

The opposite situation would be to minimize the ℓ_∞-norm of c_j. In this case, the obtained set of filters will have its energy equally distributed between all loudspeakers and therefore also along the whole spectrum, differing mainly in the phase response. As many loudspeakers are active and playing almost the same signal, the ear signals are obtained from a very intricate superposition of the many arriving wavefronts and a narrow sweet spot can be expected. Note that this formulation differs from the minimax CTC filter design proposed by RAO et al. (2007).[5]

ℓ_1 and ℓ_∞ minimization problems have, in contrast to the LSM, no analytical solution and must be solved using computationally intensive iterative methods (BOYD and VANDENBERGHE, 2004).

The frequency-domain LSM calculation is the fastest method to obtain the CTC filters and is therefore used for real-time binaural reproduction setups. The generalized frequency-domain solution given by eq. (5.9) will deliver in the shortest time the best possible channel separation for a given listener-loudspeaker setup. As ideal CTC filters are infinitely long when calculated using the DFT (KIRKEBY et al., 1998b), the obtained CTC filters might suffer from cyclic aliasing and noncausality. To minimize such artifacts, KIRKEBY et al. (1998a) proposed to extend the frequency resolution of the original transfer matrix H, which increases the calculation time, and/or to apply a regularization constraint, which adds unwanted ringing artifacts. A framework is now described for obtaining causal filters from frequency-domain calculations even when regularization is applied.

---

[4] A filter that is sparse in the frequency-domain will have an IR that is spread in time.

[5] The objective of RAO et al. (2007) is to find, in the time-domain, a set of *short* CTC filters (that per se will not ensure a perfect channel separation) whose minimum channel separation is maximized. In contrast to the frequency-domain problems described in this section, their formulation is an overdetermined problem, thus the ℓ_∞ constraint is applied to the reproduction error and not the coefficients of the CTC filter.

5.3.1 Causality

If the acoustic travel time between the loudspeakers and ears is not compensated for, the resulting CTC filters will be noncausal. This problem can be easily solved by introducing an additional latency Δ in the filters. However, regardless of this time compensation, since HRTFs are not minimum-phase, their inverse will contain a noncausal component.

Calculations of the CTC filters in the time-domain will produce a filter set which is causal, but delivers reduced channel separation in comparison to the ideal filters obtained by frequency-domain calculation eq. (5.9). If the acoustic lag is not compensated for, a noncausal filter would be required and since the product of the time-domain calculation is strictly causal, no filter would be calculated in this case. But besides the compensation of the acoustic lag, it was observed that an extra delay of approximately 1 ms will allow the presence of a certain amount of pre-ringing in the CTC filters which results in improved channel separation. This pre-ringing, originated per se from the filter calculation, will cancel out itself at the ear signal, differently than the pre-ringing originated from regularization which will remain present at the ear signal.

The frequency-domain calculation will deliver an optimal channel separation, but these filters will suffer from time aliasing. Once the filters are shifted and windowed to allow a causal response, the channel separation will also deteriorate.

To combine fast calculation time with causal filter response, a causality constraint can be imposed in the frequency-domain calculation. Using the identity

$$(\cdot)^{-1} = \mathrm{adj}(\cdot)/\det(\cdot), \qquad (5.15)$$

where $\mathrm{adj}(\cdot)$ is the adjugate of a matrix[6] and $\det(\cdot)$ its determinant, eq. (5.9) can be rewritten as

$$\boldsymbol{Y} = \boldsymbol{L} - \boldsymbol{C}D(f) = 0, \qquad (5.16)$$

where $\boldsymbol{L} = \boldsymbol{H}^* \mathrm{adj}\,(\boldsymbol{H}\boldsymbol{H}^*)\, c^{-z\Delta}$ and $D(f) = \det\,(\boldsymbol{H}\boldsymbol{H}^*)$. When written in time-domain, each element of $\boldsymbol{Y}$ is given by

$$y_{ij}(t) = l_{ij}(t) - c_{ij}(t) * d(t) = 0. \qquad (5.17)$$

---

[6]Note that for the special case of a 2×2 matrix the adjugate can be obtained without further calculation.

As discussed by PAPOULIS (1977, p. 340), a causal constraint can be applied to eq. (5.17) resulting in

$$y_{ij}(t) = l_{ij}(t) - \int_0^\infty d(t-\tau)c'_{ij}(\tau)\mathrm{d}\tau = 0. \tag{5.18}$$

Because of the causal constraint, $y_{ij}(t) = 0$ is valid only for $t > 0$. The integral in eq. (5.18) is clearly a convolution of $d(t)$ with $c'_{ij}(t)$, where $c'_{ij}(t) = 0$ for $t < 0$.

According to PAPOULIS (1977), "it suffices to find a causal function $c'_{ij}(t)$ and an anti-causal function $y_{ij}(t)$ satisfying eq. (5.18)." He does that by transforming eq. (5.18) in the Laplace-domain and arguing that $Y_{ij}(s)$ must be analytic for $\Re\{s\} < 0$ and $C'_{ij}(s)$ must be analytic for $\Re\{s\} > 0$. The transform of $l_{ij}(t)$ and $d(t)$ are uniquely determined in term of their spectra. Thus, the transform results in

$$Y_{ij}(s) = L_{ij}(-js) - C'_{ij}(s)D(-js). \tag{5.19}$$

PAPOULIS (1977) finds the solution by first factoring $D(-js)$ so that

$$D(-js) = A^+(s)A^-(s), \tag{5.20}$$

where $A^+(s)$ and its inverse $1/A^+(s)$ are analytic for $\Re\{s\} > 0$ and $A^-(s)$ and its inverse $1/A^-(s)$ are analytic for $\Re\{s\} < 0$.

The next step is to factor the ratio $L_{ij}(-js)/A^-(s)$ as the sum

$$L_{ij}(-js)/A^-(s) = B^+(s) + B^-(s), \tag{5.21}$$

where function $B^+(s)$ is analytic for $\Re\{s\} > 0$ and function $B^-(s)$ is analytic for $\Re\{s\} < 0$.

The desired causal constrained filters are then given by

$$C'_{ij}(s) = B^+(s)/A^+(s). \tag{5.22}$$

PAPOULIS (1977) concludes the solution proving that $C'_{ij}(s)$ is, as desired, analytic for $\Re\{s\} > 0$ because functions $B^+(s)$ and $1/A^+(s)$ are analytic by construction for $\Re\{s\} > 0$ and proving that $Y(s) = B^-(s)A^-(s)$ is analytic for $\Re\{s\} < 0$ because functions $B^-(s)$ and $A^-(s)$ are also analytic by construction for $\Re\{s\} < 0$.

Equation (5.22) can be rewritten in matrix form as

$$\boldsymbol{C}' = \frac{1}{\det\left(\boldsymbol{H}\boldsymbol{H}^*\right)^+} \left[\frac{\boldsymbol{H}^* \operatorname{adj}\left(\boldsymbol{H}\boldsymbol{H}^*\right)e^{-z\Delta}}{\det\left(\boldsymbol{H}\boldsymbol{H}^*\right)^-}\right]_+, \tag{5.23}$$

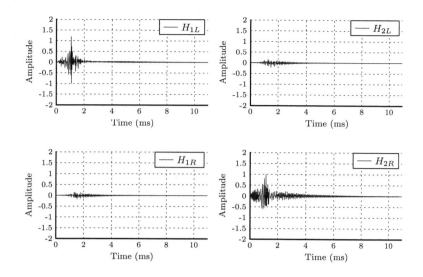

Figure 5.5: Time response of C for two loudspeakers placed at $\phi = \pm 45°$ calculated with the framework presented in section 5.3 using $\mu = 0.005$ for all frequencies and $\Delta = 3.4\,\text{ms}$. The resulting filters are strictly causal.

where $(\cdot)^+$ and $(\cdot)^-$ are, respectively, the minimum causal stable and minimum anti-causal stable parts of the determinant. As HH^* is a Hermitian matrix, $\det(HH^*)$ is real and even. In this case, the Wiener-Hopf decomposition can be efficiently implemented in the cepstral-domain allocating the first half of the cepstrum to the causal stable part and the second half for the anti-causal stable part. Further, $[\,\cdot\,]_+$ denotes the estimate of the causal part of each IR which can be obtained by windowing out the second half of the IR. KIM and WANG (2003) make the comment that as eq. (5.18) has a "linear convolution operator, not circular, care should be taken in calculating the convolution in the digital domain."

5.3.2 Regularization

Unfortunately, the transfer matrix H is not always well-conditioned, in which case the CTC filters might produce very high gains causing not only a loss of dynamic range, but also generating the so-called "ringing frequencies" (NELSON and ROSE, 2006). KIRKEBY et al. (1998a) proposed the use of regularization to limit these high gains, thus limiting the energy of the loudspeaker signal and consequently reducing loudspeaker fatigue and nonlinear behavior as well. As already discussed in section 2.2, regularization is obtained by adding a constraint on the maximum energy of the CTC filters (see appendix B.1).

The optimum filters that satisfy these constraints are given by

$$C = H^* \left(H H^* + \mu I \right)^{-1} e^{-z\Delta}. \tag{5.24}$$

The regularization parameter μ now acts as a trade-off factor between channel separation and dynamic loss. As a by-product, regularization reduces the size of the CTC filters while increasing the noncausal behavior of the filters (KIRKEBY et al., 1998a).

KIRKEBY and NELSON, 1999 showed how regularization can also be applied to calculations in the time-domain and be made frequency-dependent by filtering the *control effort* with a filter $R(z)$, resulting in

$$C = H^* \left(H H^* + \mu R(z)^* R(z) I \right)^{-1} e^{-z\Delta}, \tag{5.25}$$

where $R(z)$ attenuates all frequencies that should not be regularized. Usually, $R(z)$ has the form of a band-stop filter when it is used to design CTC filters. Note that $R(z)^* R(z)$ is real-valued and acts only as a shape-factor of the regularization that determines which frequencies are to be regularized. Assuming that the same filter will be applied to all channels, $\mu R(z)^* R(z)$ is abbreviated as $\mu(z)$ in the remainder of this thesis.

5.3.3 Minimum-Phase Regularization

Regularization introduces pre-ringing in both the CTC filters and the resulting ear signals (FIELDER, 2003; NORCROSS and BOUCHARD, 2007). As the regularization parameter is increased, the maximum amplitude of the pre-ringing component increases and the decay rate of the pre-ringing also increases.[7] This pre-ringing can result in audible artifacts if the filters are heavily regularized at certain frequencies. Since the human auditory system has a much longer post-masking behavior than pre-masking (FASTL and ZWICKER, 2007), it is desirable to alter the regularization procedure so that (at least part of) the pre-ringing is converted into post-ringing.

---

[7]It is important to stress that increasing the regularization parameter will nevertheless reduce the total filter length.

As discussed in section 2.3.2, the regularized deconvolution of a single channel can be interpreted as the direct spectral inversion multiplied by $A(z)$, the regularization shape-factor, given by

$$A(z) = \frac{1}{1 + \mu(z)/|H(z)|_2^2}, \qquad (5.26)$$

which has a real spectrum (as $\mu(z)$ is real) and, therefore, exhibit a symmetric and noncausal associated IR. NORCROSS and BOUCHARD (2007) suggest substituting $A(z)$ with its minimum-phase equivalent $A_{\mathrm{mp}}(z)$ to avoid noncausal artifacts caused by filtering the inverse of $H(z)$ with $A(z)$, and to ensure a frequency regularization without any noncausal artifacts caused by the regularization.

For the multi-channel case, the method presented in NORCROSS and BOUCHARD, 2007 has the drawback that the minimum-phase correction has to be made for each channel individually. It is possible to approximate a global minimum-phase regularization if eq. (5.15) is expanded to

$$C = \frac{\boldsymbol{H}^* \operatorname{adj}(\boldsymbol{H}\boldsymbol{H}^* + \mu(z)\boldsymbol{I})}{\det(\boldsymbol{H}\boldsymbol{H}^* + \mu(z)\boldsymbol{I})} e^{-z\Delta}. \qquad (5.27)$$

As the calculation of the adjugate of a matrix does not involve any division operation, one can assume that $\operatorname{adj}(\boldsymbol{H}\boldsymbol{H}^* + \mu(z)\boldsymbol{I}) \approx \operatorname{adj}(\boldsymbol{H}\boldsymbol{H}^*)$ as long as $\mu(z)$ is small compared to the elements of $\boldsymbol{H}$. Thus, the major influence of regularization occurs at the inversion of the determinant. Similar to eq. (2.11), the effect of regularization can be described by a regularization filter $A(z)$, so that

$$\frac{1}{\det(\boldsymbol{H}\boldsymbol{H}^* + \mu(z)\boldsymbol{I})} \equiv \frac{A(z)}{\det(\boldsymbol{H}\boldsymbol{H}^*)}, \qquad (5.28)$$

which equates to

$$A(z) = \frac{\det(\boldsymbol{H}\boldsymbol{H}^*)}{\det(\boldsymbol{H}\boldsymbol{H}^* + \mu(z)\boldsymbol{I})}. \qquad (5.29)$$

Again, as the determinant of a Hermitian matrix is real, the numerator and the denominator of eq. (5.29) will be real and thus $A(z)$ will also be real. Substituting the regularization filter $A(z)$ by its minimum-phase equivalent $A_{\mathrm{mp}}(z)$ results in

$$C_{\mathrm{mp}} = \frac{A_{\mathrm{mp}}(z)\boldsymbol{H}^* \operatorname{adj}(\boldsymbol{H}\boldsymbol{H}^* + \mu(z)\boldsymbol{I})e^{-z\Delta}}{\det(\boldsymbol{H}\boldsymbol{H}^*)}, \qquad (5.30)$$

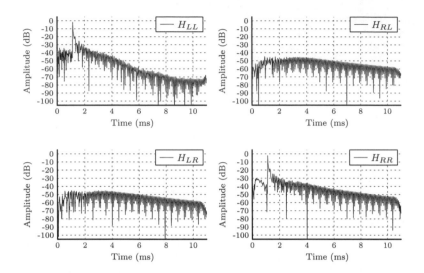

Figure 5.6: Time response of the complete transfer-path between the binaural signals and the ear signals for the filters shown in fig. 5.5. The effect of minimum-phase regularization can be observed in the impulse responses of the diagonal elements, as the impulse responses have a sharp onset (the oscillations prior to the impulse response are caused by noise, as individualized but mismatched HRTFs were used for this calculation).

which has the same amplitude response as eq. (5.25) but with all non-causal artifacts produced by regularization converted in its causal equivalent. It is also possible to combine the zero-phase with the minimum-phase of $A(z)$ in a trade-off between pre- and post-ringing (NORCROSS and BOUCHARD, 2007).

By combining eqs. (5.23) and (5.30) a causal CTC filter is obtained

$$
C'_{\mathrm{mp}} = \frac{A^+_{\mathrm{mp}}(z)}{\det\left(\boldsymbol{H}\boldsymbol{H}^*\right)^+} \left[\frac{A^-_{\mathrm{mp}}(z)\boldsymbol{H}^* \operatorname{adj}\left(\boldsymbol{H}\boldsymbol{H}^* + \mu(z)\boldsymbol{I}\right) e^{-z\Delta}}{\det\left(\boldsymbol{H}\boldsymbol{H}^*\right)^-} \right]_+ ,
$$
(5.31)

which will result in ear signals that are causal and free of pre-ringing.

5.4 Weighting

When designing a CTC reproduction system for immersive VR environments, two loudspeakers will not be sufficient to allow the listener to rotate his/her head freely. If the listener's head points in a direction outside of the arc spanned by both loudspeakers, the CTC system will become unstable (LENTZ, 2006). To meet the requirements of an immersive VR environment, LENTZ (2006) designed a system with four loudspeakers. However, as he employed the truncated CTC filter calculation algorithm (KÖRING and SCHMITZ, 1993), only two loudspeakers could be used to reproduce the binaural signals. Thus, the active pair of loudspeakers had to be exchanged according to the orientation of the listener's head. The switching between each pair of active loudspeakers was made by a soft fading between the filters. This may lead to unwanted artifacts.

To avoid such fading artifacts, all loudspeakers could be used simultaneously. On the other hand, measurements show that "two-channel configurations result in wider controlled area and are more robust to head rotation and frontal displacement than the four-channel configurations" (PARODI and RUBAK, 2010). As more sources will interact in a more complex way, smaller displacements will lead to larger errors. Thus it is reasonable to reduce the number of active loudspeakers to two,[8] but with an improved filter fading strategy.

A smoother transition between the active loudspeakers can be obtained by using a weighted matrix inversion where different weights can be applied to each loudspeaker according to the direction in which the listener's head is pointing.

The *weighted* ℓ_2 norm is given by

$$\|x\|_Z^2 = x^* Z x, \tag{5.32}$$

where Z is a diagonal matrix containing positive weights for each element of x.

The optimum set of filters that minimizes the weighted energy is given by (see appendix B.2)

$$C = W H^* \left(H W H^* + \mu(z) 1 \right)^{-1} e^{-z\Delta}, \tag{5.33}$$

---

[8]Simulation results suggest that the use of three loudspeakers will increase the robustness of the system (YANG et al., 2003).

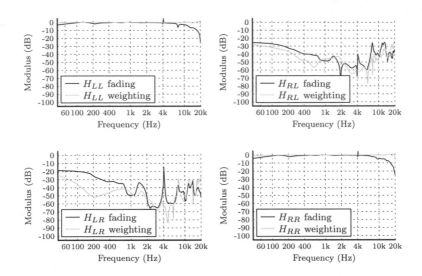

Figure 5.7: Frequency response of the complete transfer-path between the binaural signals and the ear signals using three loudspeaker for reproduction, calculated using both the fading strategy described in the work from LENTZ (2007) with the truncated CTC filter calculation algorithm and the weighting strategy presented in this work with a frequency independent regularization parameter $\mu = 0.005$. Under ideal conditions, both results should be identical.

where $\boldsymbol{W} = \boldsymbol{Z}^{-1}$. With this notation, the smaller the weight w_{ii} applied at a loudspeaker, the lower the sound pressure that this loudspeaker is supposed to generate. Thus, switching or specific fading becomes obsolete.

Applying the causality constraint and causal regularization to eq. (5.33) yields

$$C'_{\mathrm{mp}} = \frac{A^+_{\mathrm{mp}}(z)}{\det\left(\boldsymbol{HWH}^*\right)^+} \left[\frac{A^-_{\mathrm{mp}}(z)\boldsymbol{WH}^*\mathrm{adj}\left(\boldsymbol{K}\right)e^{-z\Delta}}{\det\left(\boldsymbol{HWH}^*\right)^-} \right]_+ , \qquad (5.34)$$

where $\boldsymbol{K} = \left(\boldsymbol{HWH}^* + \mu(z)\boldsymbol{I}\right)$ and $A(z)_{\mathrm{mp}}$ is the minimum-phase version of $A = \det\left(\boldsymbol{HWH}^*\right)/\det\left(\boldsymbol{K}\right)$.

5.5 Discussion

This chapter presents a general framework for the calculation of dynamic crosstalk cancellation (CTC) filters to be applied to binaural reproduction in immersive VR environments using a dynamic CTC setup with multiple loudspeakers. Such setups require high filter update rates. This means that filter calculations are performed in the frequency-domain for higher efficiency.

Since a direct calculation in frequency-domain might yield noncausal artifacts, a causality constraint in the frequency-domain calculation is introduced to avoid undesirable wrap-around effects and echo artifacts. Regularization is commonly applied to the CTC filter calculation in order to limit the output levels at the loudspeakers, which also leads, as a side effect, to noncausal artifacts. These artifacts can be minimized through the proposed minimum-phase regularization. Even though extra calculation steps are added, the calculation time required by this framework is one order of magnitude faster than an equivalent calculation in time-domain for CTC filters with 512 taps and the advantage of frequency calculation tends to increase for larger filters.

Another aspect that is especially critical for dynamic CTC systems is the switch between active loudspeakers in the setup. The use of a weighted filter calculation allows the loudspeakers' contribution to be windowed in space, resulting in a smooth filter transition free of artifacts. Weights can be made frequency-dependent, allowing for a frequency-dependent choice of active loudspeakers (cf. TAKEUCHI and NELSON, 2007).

All filter calculation described so far assumed a priori knowledge of the transmission matrix to be equalized by the CTC system. As shown in section 6.2, realistic CTC systems will not deliver a channel separation (CS) that is as high as the one obtained using an ideal CTC system. Especially at high frequencies, the obtained CS is often lower than the natural channel separation $\widehat{CS}$. GARDNER (1997, pp. 65,77) already verified this deficiency of nonindividualized CTC systems and suggested that CTC should be used only at low and middle frequencies and that the binaural signal should be played directly via two loudspeakers at high frequencies. He achieved this by bypassing the CTC filters and only equalizing the direct path between loudspeaker and ipsilateral ear. The presented framework could be expanded to include *vector base amplitude panning* (VBAP) for high frequencies, allowing the binaural signal to be smoothly panned between the loudspeakers.

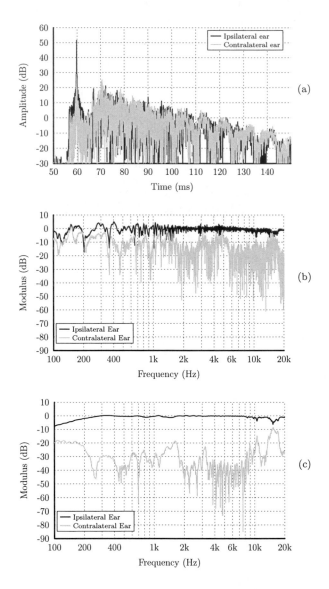

Figure 5.8: Response of a matched CTC system measured in a lightly reverberant room for the left ear of an artificial head. (a) Binaural
IR, (b) spectrum of the complete binaural IR and (c) spectrum
of the windowed binaural IR containing only the direct sound.

The framework introduced in this chapter does not take the presence of reflections in the reproduction room into account. However, in most practical applications CTC systems are built inside reverberant rooms.

The response of a matched CTC system measured in a lightly reverberant room ($T_{30} \approx 300\,\text{ms}$) can be seen in fig. 5.8. It is possible to observe in the binaural impulse response (fig. 5.8(a)) how the direct sound arriving at the contralateral ear is attenuated by over 20 dB while the room reverberation arrives at both ears with the same levels. The room reflections cause a drop in the observed CS obtained from the spectrum of the entire binaural IR (fig. 5.8(b)) when compared to the CS measured from the spectrum of the windowed binaural IR containing only the direct sound at both ears (fig. 5.8(c)).

SÆBØ (2001) studied the influence of room reflections on the localization performance delivered by a nonindividualized CTC system and concluded that room reflections can severely degrade the localization performance. Moreover, he argues that "(it could not be) shown that purely anechoic conditions are necessary (for CTC reproduction). It may well be the case that playback under nearly 'normal' conditions will be acceptable for many applications, but care and thoughtfulness should definitely be exercised."

The room reflections could be canceled by using room binaural responses instead of the HRTF in $\boldsymbol{H}$. This is, however, a more fragile process than crosstalk cancellation alone (SÆBØ, 2001). An interesting workaround to this problem is described in a publication by JUNGMANN et al. (2012) who propose a method to calculate CTC filters that are robust to (small) head displacement and room reflections. They take advantage of the masking effects of the human hearing system to design the filters so that the reflections contained in the binaural response of the system are below the masking threshold caused by the direct sound.

6

Perceptual Evaluation

This chapter presents two listening tests on two distinct aspects of individualized binaural technology. The first experiment, described in section 6.1, studies the plausibility of binaural reproduction via individually equalized headphones (BRvIEH) while the second experiment, described in section 6.2, evaluates the human sound localization performance using individualized and nonindividualized CTC systems.

Experiment I is divided into two parts: a direct and an indirect comparison of the original source, a loudspeaker, and the equivalent binaural auditory display. All 40 subjects participated in both parts of this experiment. The direct comparison was conducted as a three-alternative forced-choice test. Three stimuli were used for this test: noise, speech, and music. Results indicate that at least 50% of all listeners could not distinguish between the auditory event generated from the original source and the auditory event generated from the BRvIEH when presented with a speech or music stimulus. On the other hand, the majority of the subjects could hear a difference in the reproduction method when the presented stimulus was a pulsed pink noise. Further analysis confirmed that the observed difference in error rate between noise and the other two stimuli is significant.

An indirect comparison of the two reproduction methods was carried out in the second part of this experiment where the listeners were asked to say whether the presented stimulus originated from the headphones or one loudspeaker. In such a comparison listeners are less sensitive to differences, as no reference is provided. The pulsed pink noise was the only stimulus used for this test. Results show that no listener was able to distinguish between the original source and the BRvIEH. Furthermore, participants chose the loudspeaker more often than the headphones, which shows the authenticity of the auditory display generated using the BRvIEH.

The second experiment was aimed at testing the sound localization performance using individualized matched, individualized but mismatched, and nonindividualized crosstalk cancellation (CTC) systems.

---

§The results from experiment I were extracted from a broader study on selective auditive attention, which is described in greater depth in (OBEREM, 2012).

The individualized matched and individualized mismatched systems were based on two different sets of listener-individual HRTFs. Both sets provided similar binaural localization performance in terms of quadrant errors, polar and lateral errors, suggesting that human sound localization is robust to the HRTF measurement variations—at least to the variation levels observed when using this HRTF measurement setup. The individualized matched CTC system provided performance similar to that from the binaural listening. The localization performance deteriorated when stimuli were presented with the individualized mismatched CTC system and the errors increased even further when the nonindividualized mismatched CTC systems (based on HRTFs of other listeners) were used.

A direction-dependent analysis showed that mismatch and lack of individualization yielded a degraded performance for targets placed outside of the loudspeaker span and behind the listeners. The channel separation (CS) was also analyzed regarding its quality as a predictor for localization performance using CTC systems. The results indicate that CS might be indeed useful when it comes to evaluating mismatched CTC systems with respect to the horizontal plane localization, but a generally weak correlation was observed between the CS and the sagittal plane localization performance.

¶The virtual reality facility designed for localization tests at the Acoustics Research Institute (ARI) of the Austrian Academy of Science was used for this second experiment. The experiment was designed during a research stay of the author at ARI. To minimize costs, the HRTF measurement setup of ARI was used for this tests instead of the setup described in chapter 3.

‖The results from experiment II presented in this work have been submitted to publication in MAJDAK; MASIERO, and FELS (2012).

6.1 Experiment I: Authenticity of Binaural Reproduction via Individually Equalized Headphones

The overall quality of binaural reproduction will be influenced by aspects such as similarity of the synthesis HRTFs and the listener's own HRTFs, adequate ambient simulation, compensation of listener's head movements, and adequate sound source equalization.

Localization performance is commonly investigated as an indicator for the quality of binaural auditory displays. WIGHTMAN and KISTLER (1989) conducted listening tests comparing localization accuracy between real sources and binaural presentation. Even though error rates grew in elevation for binaural reproduction, they stated that the "appropriately synthesized stimuli presented over headphones are judged to have the same spatial positions as stimuli presented in free field." BRONKHORST (1995) also conducted listening tests on this topic. His findings showed that "virtual sound sources can be localized almost as accurately as real sources, provided that head movements can be made and that the sound is left on sufficiently long." He mentions, however, that stimuli containing considerable energy in high frequency produced poorer performance, probably caused by inadequate hardware. MØLLER et al. (1996) also conducted a similar localization test comparing the localization performance with real sources, binaural reproduction via headphones using individual HRTFs and also non individual HRTFs, concluding "that individual binaural recordings are capable of giving an authentic reproduction for which localization performance is preserved when compared to that of real life." All these studies were conducted using individually equalized headphones and, apart from one test in Bronkhorst's experiment, the stimulus was always presented in a static manner.

Localization is, however, not the only important aspect of a plausible virtual acoustic scene. An auditory event (AE) slightly displaced in relation to the original source can still generate a convincing auditory impression, as long as the AE is well externalized and no strikingly unnatural sound coloration is perceived. Thus, the authenticity of the played scene can be assumed to be a major criterion for a successful binaural reproduction and it is therefore important to examine whether the binaural reproduction can be perceptually distinguished from a real source.

The authenticity can be studied by verifying whether an AE generated by a binaural reproduction via individually equalized headphones (BRvIEH) can be distinguished from an AE generated by the original sound event. The aim of this experiment is thus to analyze if and when the BRvIEH can be distinguished from the original sound source. To achieve the best possible conditions for a plausible binaural reproduction, HRTFs and HpTFs are measured individually and listeners are required not to move their heads during the presentation. As the loudspeaker stimuli are presented to the listeners via open-type headphone, HRTFs are also measured with the listener wearing the same headphones.[1]

6.1.1 Methods

Subjects

A total of 40 listeners participated in this study. All of them stated that they have normal-hearing (no hearing test was conducted) and participated voluntarily in the experiment. All listeners were nonexpert listeners and did not receive any training to improve their localization skills. The study was performed as a blinded experiment, i.e., none of the listeners were the authors and the listeners were not enlightened as to the nature of the experiment.

HpTF and HRTF Measurements

Both HpTF and HRTF were measured individually for each listener. The HpTFs were measured eight times including a repositioning of the headphones after every measurement to allow the calculation of a robust headphone equalization filter, as described in section 4.3. An exponential sweep from 100 to 20 kHz lasting 1.73 s was used to measure each HpTF.

The HRTFs were measured at 24 loudspeaker positions (cf. section 6.1.1). The positioning of the listener's head was continuously tracked during the measurement to ensure that the listener remained still during measurement. Listeners wore the open-type headphones during the whole procedure, thus the influence of the headphones on the incoming sound field is already contained in the HRTFs.

---

[1]Even though the used headphones are of open-type, they do add a considerable coloration to the sound arriving at the listeners ears. A possible workaround would be to use custom-made tube-phones, as the ones described in (KULKARNI and COLBURN, 1998).

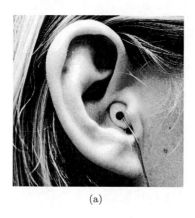 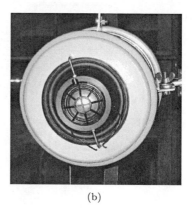

(a) (b)

Figure 6.1: Transducers used for HRTF measurement. (a) Miniature microphone fixed with ear plug in ear of participant and (b) one of the 24 custom-made coaxial loudspeakers.

The used excitation signals were exponential sweeps. The same exponential sweep used for the HpTF measurement was used, resulting in a measurement duration of less than 1 min. The influence of the transducers was removed by free-field equalization, cf. section 3.3.1.

The impulse responses of all HpTFs and HRTFs were windowed with an asymmetric Tukey window (fade out of 1 ms) to a 5 ms duration.

Microphones were fixed with an ear plug at the entrance of the listener's ear canals. To ensure a perfect fit of the microphone, the ear plug was shortened in length to be flush with the entrance of the ear canal (fig. 6.1(a)). The microphones' output signals were directly recorded via custom-made pre-amplifiers by the digital audio interface.

Apparatus and Procedure

Even though this experiment did not focus on sound localization, 24 different target positions were used to allow listeners to also use directional cues for their discrimination task. Eight loudspeakers were distributed every 45°, cf. fig. 6.3(a), along three different elevations: −30°, 0° and 30°, cf. fig. 6.3(b). All loudspeakers were placed at a distance of 2 m from the participant fig. 6.4). The target directions were randomly chosen for every new run.

The metal structure holding the loudspeakers was installed inside a fully anechoic chamber. The listener sat on a chair (with a back

Figure 6.2: Circumaural open-type headphones HD-600 from Sennheiser used
for tests presented in this section.

rest, arm rests and an adjustable head rest) placed in the middle of the
construction. To minimize the movements the head rest is adjusted for
the comfort of every single participant. To take the focus from the visual
to the aural sense, lights were turned off during the listening test (cf.
BLAUERT, 1997; MOORE, 2012).

The acoustic stimuli were generated using a computer with the ITA-
Toolbox[2] at sampling rate of 44.1 kHz and output via an external sound
card (Presonus Light pipe) connected to four analog digital converters
(ADA 8000, Behringer) with 8 channels each. Twenty-four channels are
linked to two custom-made power amplifiers, and further fed to the 24
custom-made coaxial loudspeakers, with a woofer for the frequency range
between 100 Hz to 5 kHz and a tweeter for between 3 to 20 kHz, attached
in front of the woofer (fig. 6.1(b)). Two channels are linked to the ROBO
frontend and further fed to a pair of circumaural open-type headphones
(HD-600, Sennheiser). Finally the two miniature microphones (KE-3,
Sennheiser), provided with a preamplifier each, were also connected to
AD converter.

The level of the presented stimuli was 63 dB$_{\mathrm{SPL}}$ for both headphone
and loudspeaker presentation. The stimulus' level for each presentation

---

[2]The ITA-Toolbox is a full-fledged MATLAB toolbox. The author participated
in its design and development. It is available as an open source project at
http://ita-toolbox.org/

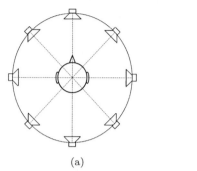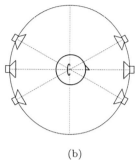

(a) (b)

Figure 6.3: Schematic distribution of loudspeakers used in experiment I distributed in (a) the horizontal plane and (b) the median plane.

Figure 6.4: Participant sitting in a chair with headrest placed in the center of the structure used to hold the loudspeakers during the listening test.

Translation	Rotation	Application
±10 mm	±2°	during measurement and playback of stimulus
±20 mm	±4°	between measurement and listening test, between input and playback

Table 6.1: Limits of tolerated movements during measurement and listening tests of experiment I.

was randomly roved within the range of ±5 dB.

A tracking sensor (Patriot, Polhemus) was mounted on the headphones' headband, which captured the position and orientation of the head in real time. The tracking data was used to monitor whether listeners remained still during measurement and signal presentation and whether they kept their head at the same position throughout all stages of the experiment. While the participant is not allowed to make greater movements than ±10 mm in translation and ±2° in rotation during the presentation of the stimulus or the measurement, the limits of movement are twice as big for the participant to find the correct position after making a decision or between measurement and listening test (cf. table 6.1).

During the test participants gave their response using a touchscreen tablet (ThinkPad Tab, Lenovo). This tablet was wirelessly connected to the main computer and acted as an extended monitor displaying a graphical user interfaces (GUI) controlled MATLAB. The participants were asked to keep the tablet on their knees during stimulus presentation and to hold it up only while entering their results. During the stimulus presentation the screen was turned off.

Test Description

Experiment I consisted of two different examinations, both with the same aim of investigating naturalness, authenticity and plausible of binaural reproduction via headphones. In the first part of this experiment a real and a synthesized stimulus were directly compared while in the second part an indirect comparison was carried out.

The whole test procedure was conducted in one section and took approximately 40 min including a break between measurements and the listening test.

Part I The first part of this experiment was a three-alternative forced choice (3-AFC) direct comparison test. The condition tested was whether a difference between a stimulus played by loudspeaker and BRvIEH could be heard. Listeners wore headphones throughout the whole test. As the loudspeaker condition was heard through the open-type headphones, the HRTFs used for synthesize the binaural signals also contained the effect of headphone attenuation (cf. section 6.1.1). Even though the attenuation of approximately 10 dB observed for frequencies above 2 kHz could influence the results, a comparison between the systems would otherwise not be possible. Before the experiment, listeners were not instructed about the kind of differences they should be paying attention to and also did not have a training phase

Three stimuli were used for this test. The first stimulus was an anechoic recording of the spoken German word "Wunschdenken" with a duration of 0.8 s. This stimulus was band limited between 200 Hz and 8 kHz. The second stimulus was a music sample, with a duration of 1.8 s. This stimulus was also band limited between 200 Hz and 10 kHz. The last stimulus was a pulsed pink noise covering the frequency range from 200 Hz to 20 kHz and with a duration of 0.8 s (each pulse had a duration of 0.3 s with a fade in and fade out of 50 ms).

Each participant listened to 20 sets of stimuli. For each set the stimulus type, reproduction combination and target direction were all chosen randomly. Furthermore, the level was roved, as explained in the previous section. After hearing a set of stimuli (up to three times), the listener had to decide which of the three presented stimuli was different than the two others and give his answer on a GUI displayed on the tablet, as shown in fig. 6.5(a).

Again, participant's head movements were observed throughout the whole test and in case they exceed the defined limits, the presented set of stimuli was considered invalid and repeated at the end of the test.

Part II In the second part of this experiment an indirect comparison test was carried out. The stimulus is reproduced either using a loudspeaker or a BRvIEH and the listeners' task is to decide whether the sound event was generated by the loudspeaker or the headphones. Part II did not include a training phase as well.

Only one stimulus type was presented in this part, namely the pulsed pink noise (described above). Every listener was presented with five stimuli via headphones and five stimuli via loudspeakers, all played in random order. Head tracking, level roving and randomized target

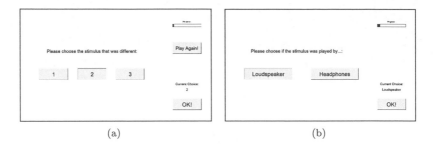

(a)   (b)

Figure 6.5: GUI design: selection menu used in experiment I for (a) the direct comparison in part I, with the instruction "please choose the stimulus that was different", and (b) the indirect comparison in part II, with the instruction " please choose if the stimulus was played by... (Loudspeaker/Headphone)".

direction are handled as in part I. After hearing each stimulus (played only once), the listener had to decide whether the stimulus had been played from a loudspeaker or from the headphones and give his answer on a GUI displayed on the tablet, as shown in fig. 6.5(b).

6.1.2 Results

Direct Comparison: Sound Quality

Prior to the analysis of the results, the collected data was analyzed for consistency. Specifically, all measured HRTF and HpTF were examined regarding abnormal behavior, e.g. dips and peaks in the lower frequency range. From this analysis one participant had to be completely excluded from the study due to an inaccurate HpTF and some single sets were excluded from other participants due to inaccurate HRTF.

The 3-AFC test consists of presenting three stimuli in random order from which two stimuli are the same and one is different. The participants are supposed to answer which of the three presented stimuli is the different one (MACMILLAN and CREELMAN, 2005). If participants were only guessing, they would have a chance out of three to choose the correct answer at each presentation, i.e., a 33.33% hit rate. In other words, a percentage of 66.67% wrong answers is expected when the subjects are guessing and therefore did not hear any difference. In case participants answered for one out of three times incorrectly (33.33%), they could

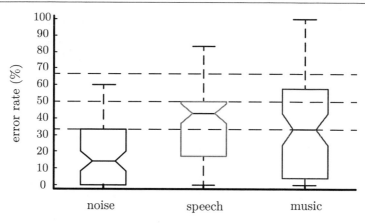

Figure 6.6: Box plot showing distribution of error rate (normalized by the total error rate of each condition) among participants in the 3-AFC discrimination test between loudspeaker reproduction and BRvIEH for three stimulus condition: noise, speech, and music.

not hear a difference for 50% of the presented stimuli. In this case, in relation to all subjects, it can be said that 50% of all listeners did not hear any difference.

Figure 6.6 shows the boxplot results for all participants and all presented stimuli. These results indicate that most of the subjects could hear a difference in the reproduction method when the stimulus was a pink noise. In numbers, 16.74% (38 out of 227) of all sets of pink noise stimuli were not answered correctly. The music stimulus presented an error rate of 35.10% (73 out of 208), slightly higher than the 33.33% limit. Therefore, at least 50% of all listeners could not distinguish between the reproduction methods. For the speech stimulus even more subjects were not able to hear any difference with an error rate of 38.17% (92 out of 241). An ANOVA with the factor condition at the three stimuli was performed. The results for music and speech were significantly different than the ones for noise ($F = 10.77$, $p < 0.001$).

The obtained data was also analyzed according to playing level and target direction. As each subject was presented with only 20 stimuli, there was in average less than two stimuli for each playing level and less than one stimulus for each target direction, making it impossible to conduct an analysis of variance. Nevertheless, simple observation of the bar chart (fig. 6.7) displaying the percentage of wrong answers (normalized by the frequency each stimulus was presented at each condition) indicates no

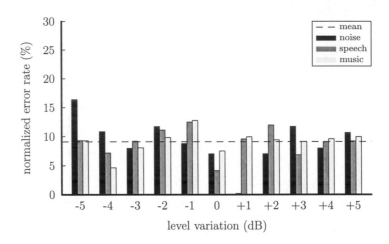

Figure 6.7: Histogram showing the distribution of errors in the 3-AFC discrimination task over different stimulus presentation level, further
subdivided into the three stimulus condition: noise, speech, and
music. The error rate was normalized by the frequency each
stimulus was presented at each condition.

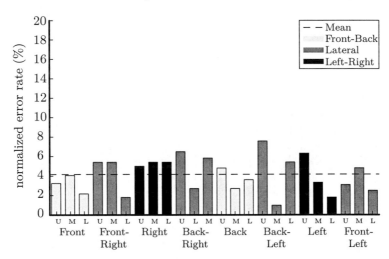

Figure 6.8: Histogram showing the distribution of errors in the 3-AFC discrimination task over different target directions for *speech* as
stimulus condition. The error rate was normalized by the frequency each stimulus was presented at each condition. U stands
for the *upper*, M the *middle* and L the *lower* loudspeakers.

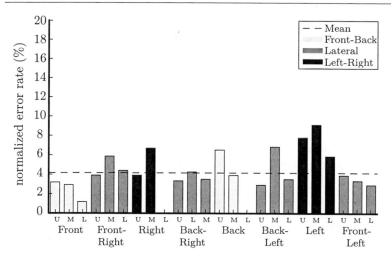

Figure 6.9: Histogram showing the distribution of errors in the 3-AFC discrimination task over different target directions for *music* as stimulus condition. The error rate was normalized by the frequency each stimulus was presented at each condition. U stands for the *upper*, M the *middle* and L the *lower* loudspeakers.

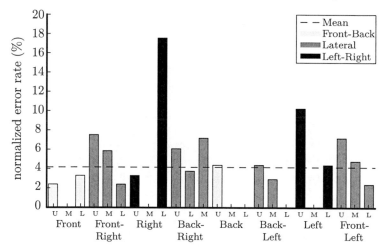

Figure 6.10: Histogram showing the distribution of errors in the 3-AFC discrimination task over different target directions for *noise* as stimulus condition. The error rate was normalized by the frequency each stimulus was presented at each condition. U stands for the *upper*, M the *middle* and L the *lower* loudspeakers.

significant difference between playing levels for all three stimuli. The same can be said about the error distribution according to the target direction, exhibited by the bar charts in figs. 6.8 to 6.10, where no significant difference between directions can be observed.

Indirect Comparison: Loudspeakers VS Headphones

Besides the direct comparison conducted in part I, an indirect comparison between loudspeaker reproduction and binaural synthesis reproduced via headphones was also made.

In a two forced-choice test as is this second part, a total error rate of 50% indicates that subjects are guessing at all times (MACMILLAN and CREELMAN, 2005). As the observed total error rate is 51.03% (199 out of 390) with a median of 50% (cf. fig. 6.11(a)), it can be assumed that no subject was able to distinguish between the reproduction methods.

Figure 6.11(b) shows that participants chose the loudspeaker (63.25%) as the reproducing method more often than the headphones (36.75%). While 32% of all stimuli presented by real sources are answered correctly, a rate of only 18% is observed for the BRvIEH.

An analysis according to the playing level was conducted. As for part I, since only 10 stimuli were presented to each participant, there is not sufficient statistical data for an analysis of variance. At a first glance, the bar chart of the total error rates indicates no significant difference in answers at each playing level.

However, an interesting result can be obtained whenever the error rate is further subdivided according to the condition reproduction method. Figure 6.12 shows the percentage of wrong choices normalized by the frequency each level was presented for each reproduction method. The chart indicates that listeners chose more often the loudspeakers as the reproduction method when the stimulus was presented with lower levels while they more often chose the headphones when the stimulus was presented with higher levels. Many listeners reported that, as no reference stimulus was presented, they did not know how to categorize the auditory event and, since they could not observe any other differences, ended up relying on the presentation level to make their decisions.

6.1.3 Discussion

In the first part of this experiment a direct comparison between loudspeaker reproduction and binaural synthesis reproduced via headphones

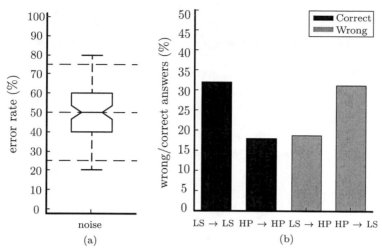

Figure 6.11: Results of the indirect discrimination task. (a) Box plot showing the distribution of the error rate among participants. (b) Histogram showing the distribution of wrong and correct answers for the four combinations of actual and perceived reproduction methods.

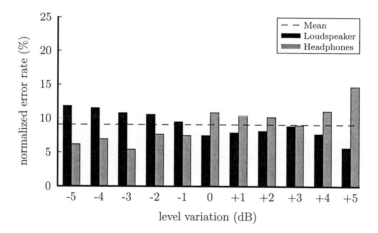

Figure 6.12: Histogram showing the distribution of errors in the indirect discrimination task over different stimulus presentation level, further subdivided into the two presentation method condition: loudspeaker and BRvIEII. The error rate was normalized by the frequency each level was presented for each reproduction method.

was also made using three different stimuli: pulsed pink noise, a speech sample, and a music sample. Results show that different numbers of subjects were able to distinguish between BRvIEH and the original sources according to the type of stimulus. This suggests that the power spectral density (PSD) of each particular stimulus plays a role in the differentiation task. While pink noise is a broad-band signal, speech is quite narrow-banded and music exhibits a PSD dominant in the range between 400 Hz and 2 kHz. It could be shown that listeners committed significantly more mistakes differentiating the reproduction methods when the played stimulus was either speech or music. This suggests that stimuli with dominant high frequency components tend to allow an easier distinction between reproduction methods.

After the listening test subjects were questioned regarding the differences they heard during the test. Most of them reported that for pink noise a different coloration was audible in higher frequencies, facilitating the discrimination task. Since headphone equalization is especially critical for frequencies higher than 4 kHz, these observations made by the listeners are reasonable.

A few subjects (2 out of 39) reported that they could hear a difference in source distance—it was not possible to verify if the sources perceived as closer were reproduced with the headphones. Reasons for this difference between BRvIEH and the original sources could be an inaccurate equalization. Further test would be required to specifically analyze the influence of headphone equalization in the perception of distance.

The most common type of difference reported by the listeners was a variation in perceived source direction of some degrees. Again, these observations are appropriate as the head displacement between measurement and reproduction could be greater than the localization blur (BLAUERT, 1997) and thus the head movement limitation implemented with the head tracking device was not strict enough. To eliminate the influence of the head displacement either the listeners' head should be fixed in a tighter manner, causing discomfort to the listener, or a dynamic binaural synthesis should be employed, a system with increased complexity and prone to new error influence.

In the second part of this experiment an indirect comparison between loudspeaker reproduction and binaural synthesis reproduced via headphones was made. This test was conducted with only one stimulus: the pulsed pink noise, which is expected to be the stimulus that will make listeners most sensitive to the differences between the two reproduction methods. However, results showed that not a single listener was able to consistently identify whether the auditory event was generated by a loudspeaker or by the headphones.

Since subjects were not able to find differences for this stimulus, it can be assumed that subjects will also not be able to distinguish between real sources and BRvIEH for stimuli like music and speech.

The fact that the listeners could not tell these two reproduction methods apart indicates that the BRvIEH sounded natural and authentic, i.e., even though differences between the two methods can be heard when stimuli possessing significant content in high frequencies are played, these differences are not big enough to allow listener to perceive that the auditory event binaurally reproduced via individually equalized headphones is actually coming from the headphones and not from the external source.

6.2 Experiment II:
Localization Performance with Individual-ized and Nonindividualized CTC Systems

As discussed in chapter 5, CTC filters are calculated based on the transfer paths between loudspeakers and listener's ears, i.e., the HRTFs. So far, it has been assumed that exactly the same HRTFs are used for the filter calculation and the listening situation, a so-called *matched* CTC system, which provides optimal crosstalk cancellation. In a *mismatched* CTC system, the HRTFs do not exactly match the CTC filters and the performance is assumed to degrade. The actual localization performance of a CTC system has already been investigated in the horizontal plane (GARDNER, 1997; TAKEUCHI et al., 2001; BAI and LEE, 2006; LENTZ, 2006), however, little is known about the localization performance in *both* horizontal and sagittal planes provided by a CTC system.

AKEROYD et al. (2007) used HRTFs from other listeners to create mismatched CTC systems and compared their numeric performance, based on the obtained channel separation (CS) with the matched CTC systems. In a simulation of binaural processing, they showed disrupted ITDs and ILDs for the mismatched CTC systems. Through their simulation results they concluded that the mismatched system will probably yield a degraded lateral localization performance, particularly for directions with a high value for the lateral angle α.

Even though AKEROYD et al. (2007) used listener-individual HRTFs to create a matched CTC system, the listener-individual HRTFs do not always yield a matched CTC system. For example, if the HRTF measurement is repeated for the same listener, the HRTF set will still be considered as listener-individual, but acoustic properties of the HRTFs would slightly change, causing a mismatch to the CTC filters. This is actually a common situation, even in individualized CTC systems, where the propagation paths change between the HRTF measurements and the actual use of the CTC system.

Thus, the aim of this listening test is to investigate two-dimensional human localization performance in CTC systems with a special focus on individualized matched, individualized but mismatched, and nonindividualized CTC systems. The individualized but mismatched CTC systems used a second HRTF measurement of the same listeners. The nonindividualized CTC systems used HRTFs from a mannequin and other listeners. Also, the baseline performance was acquired for binaural sound presentation without any CTC filtering.

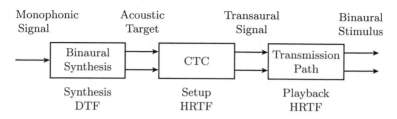

Figure 6.13: Block diagram for the signal processing conducted for the preparation of experiment II. A monophonic signal is first filtered by the individual DTF at the binaural synthesis stage. The resulting acoustic target is further filtered by the CTC filter to generate the transaural signals, which are further filtered by the individual HRTFs corresponding to two loudspeakers, resulting in the binaural signals presented to the listener.

Channel separation (see section 5.2) is commonly used to describe the quality of a CTC system (GARDNER, 1997; BAI and LEE, 2006). AKEROYD et al. (2007) showed that much smaller CSs are obtained in mismatched CTC systems compared with matched CTC systems. Recently, PARODI and RUBAK (2011) investigated the minimum audible channel separation in an artificial CTC system. However, it is still not clear how the channel separation is related to the localization performance. Thus, a comparison between the channel separation values and the sound-localization performance in CTC systems is conducted to investigate its use as a predictor for the localization performance.

The channel separation, being a frequency-dependent measure, is usually averaged over a frequency range in order to describe a CTC system by a single value. Keeping in mind that different frequency regions contribute differently to the sound localization in the horizontal and sagittal planes, as discussed in section 2.4.2, it was investigated whether channel separation calculated in specific frequency regions better describes different aspects of the sound localization.

Current CTC systems usually suffer from various technical limitations (cf. chapter 5). The cancellation quality depends to a large extent on the listener's alignment between the loudspeakers (TAKEUCHI et al., 2001) and loudspeaker combinations have been proposed to increase the area of the sweet spot (TAKEUCHI and NELSON, 2007). Loudspeakers, usually simulated as point sources, have non-ideal transfer function and directionality, which also have a strong effect on the quality of the CTC systems (QIU et al., 2009). Other potential artifacts are head movements and room reflections.

In order to better control the issues listed above, the localization performance was tested using a *virtual CTC system*. Thus, a binaural simulation of the CTC system (see fig. 6.13) was used, i.e., the stimulus was presented via headphones and individualized HRTFs were used to simulate the propagation paths between loudspeakers and the listener. The virtual CTC system consisted of three different filter stages: 1) listener-individual DTFs, used to create an acoustic target; 2) CTC filters, used to create the transaural signals for the virtual loudspeakers, and; 3) listener-individual HRTFs, used to simulate the virtual loudspeakers. In such a setup, the loudspeaker effects were reduced to that of the HRTF measurement and the listeners' head was always virtually fixed within the sweet spot.

6.2.1 Methods

Subjects

Eight listeners participated in this study,[3] all of them having absolute hearing thresholds within the 20 dB range of the average normal-hearing population in the frequency range between 0.125 and 12.5 kHz. All listeners who participated in this test had participated in previous test and had therefore previous experience with localization tests. They all showed front-back confusion rates below 20% in pre-experiments with their own broadband DTFs. As experiment I, this study was also performed as a blinded experiment. All participants received a financial compensation for taking part in this listening test.

HRTF Measurements

HRTFs were measured individually for each listener. The measurement setup at ARI is also composed of a supporting arc, containing 22 loud-speakers (custom-made boxes with VIFA 10 BGS as drivers) at fixed elevations from -30° to 80°, with a 10° spacing between 70° and 80° and 5° spacing elsewhere. The listener was seated in the center point of the circular arc on a computer-controlled rotating chair. The distance between the center point and each speaker is 1.2 m. Miniature

---

[3]Compared to exp. I, the number of subjects in exp. II might seem insufficient for an adequate statistical analysis. However, in exp. I each listener was tested for only one condition and did relatively few repetitions. Therefore, many listeners had to be pooled together. Exp. II compares the performance of each individual listener, conducting many repetitions for each condition. Thus, in this case, eight subjects can be considered a satisfactory sample size.

microphones (Sennheiser KE-4-211-2) were inserted into the listener's ear canals and their output signals were directly recorded via amplifiers (FP-MP1, RDL) by the digital audio interface.

The used excitation signal was a multiple exponential sweep, cf. section 3.2.1. A 1.73 s exponential frequency sweep from 0.05 to 20 kHz was used to measure each HRTF. At an elevation of $\theta = 0°$, the HRTFs were measured with a horizontal spacing of $\Delta\phi = 2.5°$ within the range of $\phi = \pm 45°$ and with the horizontal spacing of $\Delta\Phi = 5°$ otherwise. According to this rule, the measurement positions for other elevations were distributed with a constant spatial angle, i.e., the azimuthal spacing increased towards the poles. In total, HRTFs for 1550 positions within the full 360° horizontal span were measured for each listener. The measurement procedure lasted approximately 20 minutes. As described in section 3.3.1, first the influence of the transducers was removed by equalizing the HRTFs. Then the directional transfer functions (DTFs) were calculated. Finally, the impulse responses of all HRTFs and DTFs were windowed with an asymmetric Tukey window (fade in of 0.5 ms and fade out of 1 ms) to a 5.33 ms duration.

Two sets of HRTFs were measured for each listener.[4] The first measurements were performed for a previous study—all current participants took part at this previous study—and the second measurements were performed for the present study. The interval between the two measurements was approximately five years.

Acoustic Targets

Lateral and polar angles from the horizontal-polar coordinate system (see fig. 2.6) were used to describe the acoustic target's position (MORIMOTO and AOKATA, 1984). The tested lateral angle ranged from $\alpha = -90°$ (right) to $\alpha = 90°$ (left). The polar angle of the targets ranged from $\beta = -30°$ (front, below eye-level) to $\beta = 210°$ (rear, below eye-level). The targets were pseudo-uniformly distributed on the surface of the sphere by using a uniform distribution for the polar angle and an arcsine-scaled uniform distribution for the lateral angle.

The acoustic targets were Gaussian white noises with a duration of 500 ms and 10 ms fade-in and fade-out, filtered with the listener-specific DTFs. Prior to filtering, the position of the acoustic target was discretized to the grid of the available DTFs.

---

[4]The measured HRTFs and DTFs are available at http://www.kfs.oeaw.ac.at/ hrtf. The listeners are referred throughout the work by the same anonymous identification number used on the online database. NH stands for *normal hearing*.

The level of the presented stimuli was 50 dB above the individual absolute hearing threshold in each condition. The threshold was estimated in a manual up-down procedure individually for each condition using an acoustic target positioned at lateral and polar angle of 0°. As in experiment I, the stimulus level for each presentation was randomly varied within the range of ±5 dB to reduce the possibility of localizing spatial positions based on overall level.

Binaural CTC Simulation

In the tested CTC conditions (see section 6.2.1), the acoustic targets were processed with a binaural CTC simulation. The simulation was used to ensure that subjects were always in the sweet-spot and to fully control the correspondence between the acoustic paths and CTC filters.

The CTC filters were calculated for a pair of virtual loudspeakers with one loudspeaker placed at $\phi = 45°$ left and second loudspeaker placed at $\phi = -45°$ right to the listener, both at $\theta = 0°$. Thus, the loudspeaker span angle was $\Delta\phi = 90°$.

The propagation paths from the loudspeakers to the listener's ears are described by the so-called "setup HRTFs". The corresponding impulse responses were zero padded to 85.33 ms.[5] The CTC filters were calculated in the frequency-domain according to eq. (5.24) with $\beta = 0.005$ for all frequencies.[6] The CTC filters were converted back to the time-domain and circularly shifted by 3.125 ms to avoid noncausality. Finally, the impulse responses where windowed with a one-sided Tukey window with a fade out of 18.6 ms at their end.

The transaural signals were calculated by processing the acoustic target with the CTC network according to eq. (5.4). Then, the transmission of the transaural signals from the loudspeakers to the listener's ears was simulated by filtering the transaural signals using the listener-individual HRTFs, the so-called "playback HRTFs". Note that listener-individual HRTFs were used for the playback HRTFs in all conditions—only the setup HRTFs were varied in this study.

---

[5]Note that 85.33 ms correspond to 4096 samples. All the signal processing calculations in this experiment were done at the sampling rate of 48 kHz.

[6]To allow comparison of this experiment's results with the results from AKEROYD et al. (2007), the CTC filters here were calculated in the same way as described by them. Therefore, causality constraint (section 5.3.1) and minimum-phase regularization (section 5.3.3) were not used.

Apparatus and Procedure

The virtual acoustic stimuli were presented via headphones (HD 580, Sennheiser) in a double-wall sound-proof room. The headphones were diffuse-field-compensated circumaural headphones and no additional headphone correction was applied,[7] as DTFs were used for the binaural synthesis (LARCHER et al., 1998). The listener stood on a platform enclosed by a circular railing. Stimuli were generated using a computer and output via a digital audio interface (ADI-8, RME) with a 48 kHz sampling rate. A virtual visual environment was presented via a head-mounted display (3-Scope, Trivisio). It provided two screens with a field of view of $32° × 24°$ (horizontal × vertical dimensions). The virtual visual environment was presented binocularly with the same picture for both eyes. A tracking sensor (Flock of Birds, Ascension) was mounted on the top of the listeners' head, which captured the position and orientation of the head in real time. A second tracking sensor was mounted on a manual pointer. The tracking data were used for the 3-D graphic rendering and response acquisition.

The listeners were immersed in a spherical virtual visual environment (MAJDAK et al., 2010). They held a pointer in their right hand. The projection of the pointer direction on the sphere's surface, calculated based on the position and orientation of the tracker sensors, was visualized and recorded as the perceived target position. The pointer was visualized whenever it was in the listeners' field of view.

Prior to the tests, listeners performed a visual and an acoustic training. The aim of the visual training was to train subjects to perform accurately in the virtual environment. The visual training was a simplified game in the first-person perspective where listeners had to find a visual target, point at it, and click a button within a limited time period. This training was continued until 95% of the targets were found with a root-mean-square (RMS) angular error in the range of 2°. This performance was achieved within a few hundred trials. Then the acoustic training was performed with listener-individual DTF (MAJDAK et al., 2010). The goal of the acoustic training was to ensure a stable localization performance of the subjects. The acoustic training consisted of 6 blocks, 50 acoustic targets each, lasting approximately 2 hours.

---

[7]According to SCHONSTEIN et al. (2008), the impact of the headphone equalization on the binaural localization performance is still arguable. Unfortunately, their test was conducted with only one subject—who also designed the test—therefore lacking in statistical significance.

Figure 6.14: Overview of the anechoic chamber at ARI. In the front-plane the
HRTF Measurement arc with its 22 loudspeakers and computer-
controlled rotating chair. In the background the platform where
the localization tests were conducted. A listener wearing the
head-mounted displays holds the pointing device. The position
of the listener's head and pointing device are tracked by a
tracking device whose sender is placed in front of the test
platform.

In the actual acoustic tests, at the beginning of each trial, the
listeners were asked to align themselves with the reference position and
click a button. Only after that the stimulus was presented. During the
presentation, the listeners were instructed not to move, cf. section 2.4.2.
The listeners were asked to point to the perceived stimulus location and
click the button again. This response was recorded for the data analysis.
The tests were performed in blocks; each block consisted of 100 acoustic
targets and took approximately 15 minutes. Within a block, the targets
are first randomly selected (cf. section 6.2.1) and then sampled to the
nearest neighbor from the 1550 possible spatial positions. After each
block, subjects had a break of approximately 15 minutes. The procedure
was controlled by LocaCTC from the ExpSuite.[8]

---

[8]Available at http://sf.net/projects/expsuite.

Conditions

Eight conditions were tested in three blocks each. The order of the blocks was randomized in such a way that within eight blocks all conditions were in a randomized order.

The first two conditions consisted of pure acoustic targets, i.e., binaural signals without the CTC simulation. The former, *binOwn*, used the same DTFs as those used for the acoustic training while the latter, *binOwnB*, used the more recently measured DTFs.

In the individual matched CTC condition, *ctcOwn*, the acoustic targets were presented via the simulated CTC system using the same setup and playback HRTFs, namely the listener-individual HRTFs from the condition binOwn. The condition ctcOwn corresponds to the matched case from AKEROYD et al. (2007) and represents an *ideal individualized* CTC system where the CTC filters match exactly the acoustic paths between the loudspeakers and the listener.

In the individual but mismatched CTC condition, *ctcOwnB*, the playback HRTFs were the same as in the matched CTC condition, while the setup HRTFs were those from the more recent measurement, corresponding to the DTFs used for condition binOwnB. The condition ctcOwnB represents a *realistic individualized* CTC system where for the calculation of the CTC filters, the listener-individual HRTFs have been measured, but during the signal presentation, the acoustic propagation paths do not exactly match these measured HRTFs. Note that from an acoustic point of view, this condition is a mismatched condition.

The last CTC conditions were *nonindividual* mismatched conditions, i.e., the setup HRTFs were those from other sources, while the playback HRTFs did not change. In the condition *ctcKemar*, the setup HRTFs were those from measurements on a mannequin (GARDNER and MARTIN, 1995). Note that in contrast to all other HRTFs used in study, the mannequin's HRTFs were measured using microphones included in an ear simulator, yielding an HRTF set containing the direction independent ear-canal transfer function. In the remaining nonindividual conditions, the setup HRTFs were those from other listeners, namely, NH57, NH64, and NH68. These particular listeners were also tested with setup HRTFs from NH12 in order to obtain the same number of tested conditions for each listener. Those conditions are referred to as *ctcNH57*, *ctcNH64*, *ctcNH68*, and *ctcNH12*. For the sake of simplicity, all nonindividual conditions are referred to as *ctcOther*.

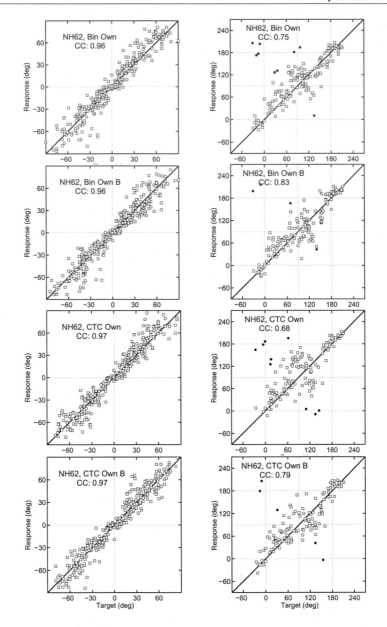

Figure 6.15: Localization results of the exemplary listener NH62 for all
tested conditions. Lateral results are plotted in the left panels, the polar results in the right panels. Polar results outside

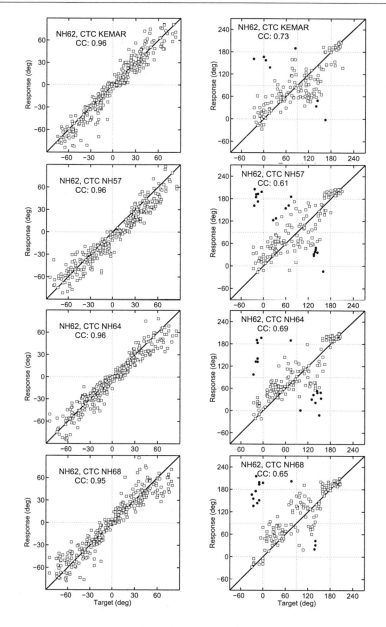

Figure 6.15: (cont.) the lateral range of $\pm\,30°$ are not shown. Filled circles: Responses with errors outside the $\pm\,90°$ range. CC: Correlation coefficient between responses and targets.

6.2.2 Results

Localization Performance: Binaural Reproduction

Figure 6.15 shows results of the localization experiment for an exemplary listener (NH64). The target and response angles are shown on the horizontal and vertical axes, respectively, of each panel. For the polar dimension, the results are shown for targets with lateral angles within $\pm 30°$ only. Responses that resulted in absolute polar errors larger than $90°$, i.e. can be considered as quadrant errors, are plotted as filled circles. All other responses are plotted as open squares. The performance seems to be similar for both binaural conditions and the differences to the ctcOwn condition seem to be negligible. A generally degraded performance can be observed for ctcOwnB and also all other mismatched conditions.

As defined in section 2.4.2, localization errors were calculated by subtracting the target angles from the response angles. The lateral error (LE) is used to measure localization performance in the horizontal plane. In the polar dimension, data is analyzed in regard to confusions between the hemifields, measured with the quadrant error (QE), and the local performance within the correct hemifield, measured with the polar error (PE). Only responses within the lateral range of $\pm 30°$ were considered in the polar dimension analysis (MIDDLEBROOKS, 1999b).

The results described by the error metrics QE, PE, and LE are shown in tables 6.2, 6.3 and 6.4, respectively. In both binaural conditions, the group average performance was within the range of previously reported performance for localization of virtual broadband noises under comparable conditions. For the sagittal planes, the average QE of 8.6% (for binOwn, 7.1% for binOwnB) was similar to QE of 7.7% from MIDDLEBROOKS (1999b) and to QE of 9.4% from GOUPELL et al. (2010). Also, the average PE of 31.0° (31.6° for binOwnB) was similar to PE of 28.7° from MIDDLEBROOKS (1999b) and to PE of 33.8° from GOUPELL et al. (2010). For the horizontal planes, the average LE of 10.7° (10.4° for binOwnB) was similar to LE of 14.5° from MIDDLEBROOKS (1999b) and to LE of 12.4° from MAJDAK et al. (2011).

Repeated measures (RM) analysis of variance (ANOVA) was used for the statistical analysis of the results. Each of the three tested blocks was treated as within-subject repetition. For the binaural conditions, RM ANOVAs were calculated on the LE, PE, and QE with the factor condition at two levels (binOwn, binOwnB). The analysis showed neither significant effect for the QE ($p = 0.44$), nor for PE ($p = 0.68$), nor for

LE (p = 0.38). This indicates that despite of five years of break between the two HRTFs measurements, both HRTF sets provided localization performance at a similar level.

Localization Performance: Individualized CTC systems

In order to investigate the performance in individualized CTC systems, RM ANOVAs with the factor condition at four levels (binOwn, binOwnB, ctcOwn, and ctcOwnB) were performed. The results were significant for the QE (p = 0.005) and for the LE (p < 0.001), but not for the PE (p = 0.20). Tukey-Kramer *post-hoc* tests were used to test the statistical significance of particular levels. The significance was considered at p < 0.05. Post-hoc tests showed that only the ctcOwnB condition yielded significantly larger QE and LE compared to all other conditions. Note that even though for PE the differences were not significant, the PE was larger for ctcOwnB (34.2°) than for ctcOwn (31.8°) or the binaural conditions (31.0° and 31.6°).

The lack of significance in the differences between the ideal CTC and binaural systems indicates that the ideal CTC system used in this test provided localization performance at the level of the binaural reproduction systems. In a realistic application of the individualized CTC, this situation is, however, unachievable because the propagation paths would (slightly) change as soon as a listener leaves the HRTF measurement setup and enters the CTC system. This situation was represented by the condition ctcOwnB, where individual HRTFs from the latter measurement were used to calculate CTC filters. Note that this is not a *worst-case scenario* as head movements may induce stronger changes to the actual playback HRTF. The performance in such a realistic CTC system was worse than that in an ideal CTC system in terms of significantly larger QEs and LEs. This demonstrates that a mismatch between the playback and setup HRTFs may result in a degraded localization performance in a CTC system, even when both HRTFs provide a similar performance in a binaural system.

Compared to ctcOwnB, the ctcOwn condition yielded a better performance in the horizontal plane. This result confirms the results for modeling interaural differences in matched and mismatched CTC systems (AKEROYD et al., 2007) where for mismatched CTC systems, the model predicted large ITD and ILD errors.

Condition	NH12	NH14	NH15	NH57	NH62	NH64	NH68	NH72	Mean
binOwn	3.3	5.3	7.4	25.4	5.9	2.7	6.3	12.5	8.6
binOwnB	1.7	5.2	1.5	17.5	2.8	6.1	13.3	9.1	7.1
ctcOwn	4.2	7.4	6.8	15.2	6.7	1.4	13.5	8.9	8.0
ctcOwnB	0.6	6.3	5.0	33.0	4.0	11.0	32.3	16.7	13.6
ctcOther	6.3	9.7	16.3	25.8	10.0	13.2	33.2	20.7	16.9
ctcKemar	2.0	7.8	10.1	23.1	5.4	12.2	36.6	5.1	12.8
ctcNH57	10.6	11.5	14.2	-	11.8	4.4	36.6	23.9	16.1
ctcNH64	0.0	4.4	18.5	28.3	10.8	-	29.8	23.2	16.4
ctcNH68	21.9	12.1	27.1	23.3	9.1	23.1	-	18.3	19.3
ctcNH12	-	-	-	28.3	-	14.2	25.3	-	22.6

Table 6.2: Quadrant errors (QE) in % for all listeners and conditions tested. The condition ctcOther represents the median of the nonindividual conditions.

Condition	NH12	NH14	NH15	NH57	NH62	NH64	NH68	NH72	Mean
binOwn	28.3	26.5	30.5	37.0	27.0	28.5	31.1	38.9	31.0
binOwnB	25.0	30.1	31.2	35.6	26.6	34.8	34.2	35.5	31.6
ctcOwn	26.7	25.8	36.2	33.9	35.0	32.0	26.7	38.4	31.8
ctcOwnB	26.8	35.6	32.2	36.7	29.0	32.0	41.2	39.8	34.2
ctcOther	35.4	31.9	40.1	39.4	34.1	33.7	37.4	41.5	36.7
ctcKemar	35.2	26.0	39.9	33.7	35.2	31.3	40.6	36.8	34.8
ctcNH57	35.5	31.4	40.4	-	34.4	33.5	42.9	42.7	37.2
ctcNH64	26.1	32.3	36.0	42.2	31.3	-	32.3	43.6	34.8
ctcNH68	36.8	32.8	41.1	38.2	33.7	33.9	-	40.4	36.7
ctcNH12	-	-	-	40.6	-	38.0	34.2	-	37.6

Table 6.3: Local polar error (PE) in degrees for all listeners and conditions tested. The condition ctcOther represents the median of the nonindividual conditions.

Condition	NH12	NH14	NH15	NH57	NH62	NH64	NH68	NH72	Mean
binOwn	8.1	10.1	10.9	16.7	9.7	9.4	10.0	10.7	10.7
binOwnB	8.2	9.6	12.7	13.6	9.9	9.2	9.4	10.6	10.4
ctcOwn	8.0	9.1	11.7	14.0	9.6	8.7	10.4	11.1	10.3
ctcOwnB	10.8	13.4	14.3	15.5	9.5	10.2	11.8	18.2	13.0
ctcOther	11.0	11.3	14.4	15.5	10.9	10.8	13.0	14.0	12.6
ctcKemar	9.8	9.8	14.1	14.8	9.4	11.3	13.7	9.3	11.5
ctcNH57	11.9	8.9	14.8	-	12.8	10.2	13.6	14.4	12.4
ctcNH64	10.0	12.8	13.7	16.2	10.4	-	12.3	13.5	12.7
ctcNH68	13.0	16.2	17.2	17.6	11.4	15.8	-	14.6	15.1
ctcNH12	-	-	-	14.7	-	10.0	11.4	-	12.0

Table 6.4: Lateral error (LE) in degrees for all listeners and conditions tested. The condition ctcOther represents the median of the nonindividual conditions.

Looking more closely at the simulation results presented in (AKEROYD et al., 2007, fig. 10; reproduced in fig. 6.16), it seems that the errors are large for large negative interaural differences only. For smaller interaural differences, the errors seem to be negligible, which would suggest a correct reproduction of central targets.

In order to investigate this issue, the targets were grouped to those within (central) and those outside (lateral) the loudspeaker span and the LEs were calculated as a function of the target lateral angle (fig. 6.17(a)). While in the ctcOwnB condition the performance seems to slightly degrade for the central targets, the performance appears to be much worse for the lateral targets.

An RM ANOVA was performed on the LEs for the factors target direction (central, lateral) and condition (ctcOwn, ctcOwnB). Both main effects ($p < 0.001$) and their interaction ($p = 0.048$) were significant. The significant interaction suggests a different impact of the condition for the two target directions. The post-hoc test showed that the only significant difference was that for the lateral targets tested with ctcOwnB ($17.2°$) when compared to the lateral targets tested with ctcOwn ($12.5°$) or when compared to the central targets ($11.5°$ for ctcOwnB and $9.7°$ for ctcOwn).

This indicates that while for targets placed outside the loudspeaker span the mismatch significantly affects the lateral localization performance, for targets placed inside the span the mismatched CTC system

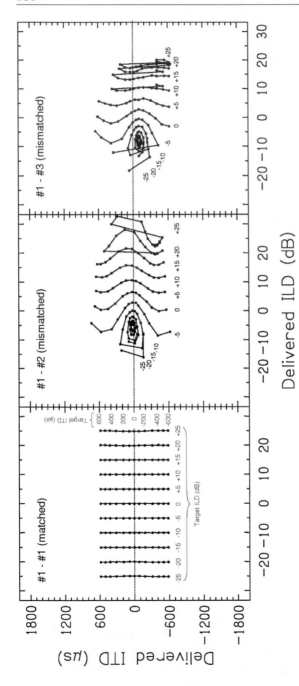

Figure 6.16: The binaural performance of a virtual CTC system, calculated for a matched system (left panel) and two mismatched systems (middle and right panels). Each panel shows the ongoing ITD (ordinate) and ILD (abscissa) delivered by the simulation for a large set of combinations of target ITD and ILDs (parameters); the lines join points with the same target ILD. The analysis was run at an auditory-filter frequency of 1000 Hz.

Reprinted with permission from M. A. AKEROYD et al. (2007). "The binaural performance of a cross-talk cancellation system with matched or mismatched setup and playback acoustics". In: *J. Acoust. Soc. Am.* 121.2, pp. 1056–1069. Copyright 2007, Acoustical Society of America.

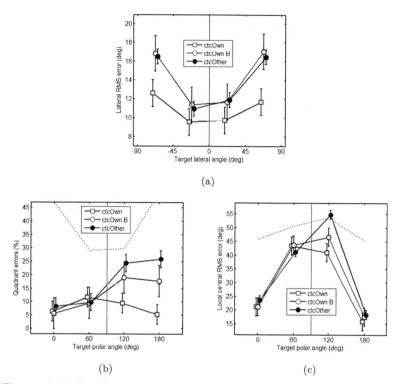

Figure 6.17: Localization error as a function of target position. (a) Lateral error as a function of the lateral angle. (b) Quadrant errors and (c) polar errors as functions of the polar target angle. The dashed lines show the errors which would result from random responses.

may yield a similar performance as the matched CTC system. This seems to confirm the above-mentioned observations concerning the details of binaural modeling described in AKEROYD et al. (2007). It could be speculated that there exist a correspondence between central targets in a mismatched CTC system and phantom sources in a stereophonic reproduction system.

Targets placed at elevations near the loudspeakers may also correspond to a phantom-source stereophonic reproduction. If such targets were well-localized in sagittal planes even in a mismatched CTC system then the difference between ctcOwn and ctcOwnB would depend on the target polar angle, with a larger difference in the performance for targets placed behind the listener.

Figure 6.17(b) shows the QEs as a function of the polar angle, with targets grouped to four groups with a polar angle span of $60°$ (starting at $\beta = -30°$). The QE seems to increase with the polar distance between the targets and the loudspeaker elevation.

An RM ANOVA was performed with factors target hemifield (frontal targets: $-30° \leq \beta \leq 30°$, rear targets: $150° \leq \beta \leq 210°$) and condition (ctcOwn, ctcOwnB). The main factors ($p < 0.018$ for both) and their interaction ($p = 0.023$) were significant. The significant interaction suggests a different impact of the conditions in the two target hemifields. The post-hoc test showed that while for the frontal targets, the difference between the conditions was not significant (5.5% for ctcOwn, 5.9% for ctcOwnB), for the rear targets, highly-significantly ($p < 0.005$) more QEs occurred in the ctcOwnB (16.8%) than in the ctcOwn (5.7%) condition. This indicates a strong impact of the mismatch on the localization performance for the targets placed behind the listener. This is in agreement with the result presented by NELSON et al. (1997), who conducted a localization test using a mismatched CTC system using real loudspeakers and verified that when the system worked poorly, the front-back inversion rate (thus, the quadrant error) increased substantially.

All in all this means that, for the targets placed in the same hemisphere as the loudspeakers, the individual but mismatched condition ctcOwnB yielded a sagittal plane performance similar to that for the ideal matched condition ctcOwn. For the targets placed in the opposite hemisphere as the loudspeakers, the QE were substantially larger. This indicates that mismatched but individualized CTC systems might be able to provide a good performance only for the frontal targets. The exact match of the CTC filters to the propagation paths seems to be highly relevant for targets virtually placed at the other hemisphere than the loudspeakers. Furthermore, results indicate that only an ideal, individualized, matched CTC system provides a correct reproduction of the spectral cues required for accurate sagittal plane sound localization in both hemifields.

Localization Performance: nonindividualized CTC systems

The localization performance in the nonindividualized and thus mismatched CTC systems is usually assumed to be worse than that in individualized CTC systems (AKEROYD et al., 2007). However, an individualized CTC system can also be matched or mismatched, depending on whether an ideal or a realistic CTC system is being studied. Thus, in the following the nonindividualized CTC systems (ctcOther) are compared with both the ideal CTC system (ctcOwn) and the realistic CTC system (ctcOwnB).[9]

The LE increased from 10.3° (ctcOwn) to 13.0° (ctcOwnB), but then it decreased to 12.6° (ctcOther), indicating a weak impact of the individualization on the horizontal plane localization performance (see table 6.4). However, there might have been some differences at particular target directions. Thus, the targets were grouped to those within (central) and those outside (lateral) the loudspeaker span and the LE were calculated as a function of the target lateral angle (fig. 6.17(a)). LE was still similar for all the mismatched (individual and nonindividual) conditions. Thus, for the horizontal plane localization, there seems to be no difference between the two types of mismatched CTC systems and an individualized CTC system seems to be of no advantage.

The QE increased from 8.0% (ctcOwn) to 13.6% (ctcOwnB) and then further to 16.9% (ctcOther, see table 6.2). Also, the PE increased from 31.8° (ctcOwn) to 34.2° (ctcOwnB) and then further to 36.7° (ctcOther, see table 6.3). The RM ANOVAs were performed with the factor condition at three levels (ctcOwn, ctcOwnB, and ctcOther) on the QEs and PEs. The factor condition significantly affected the QEs ($p < 0.001$) and the PEs ($p = 0.019$). The post-hoc tests showed that while ctcOther yielded significantly larger errors compared with ctcOwn, the errors were not significantly different when compared with ctcOwnB. At first glance, this might indicate that for the sagittal plane localization the performance is independent of the individualization of the filters. However, differences at particular positions might have been expected as was the case for the individualized mismatched condition. Thus, the targets were grouped into four groups and the QE and PE were calculated as a function of the polar angle (figs. 6.17(b) and 6.17(c)). For ctcOther, the QE increased with the increasing distance between the targets and the loudspeakers more than it did for ctcOwnB.

An RM ANOVA with factors target hemifield (frontal targets: -30° to 30°, rear targets: 150° to 210°) and condition (ctcOwnB and ctcOther) was performed on the QEs. The main factors condition ($p = 0.048$) and target hemifield ($p < 0.001$) were significant, but their interaction was not ($p = 0.23$). For the front targets, the QE was 5.8% and 8.1% for ctcOwnB and ctcOther, respectively. For the rear targets, the QE was

---

[9]The performance also varied across particular nonindividual conditions (see ctcKE-MAR versus ctcNH68). However, in order to increase the statistical power, all nonindividual CTC conditions were pooled together in the presented analysis.

16.8% and 26.1% for ctcOwnB and ctcOther, respectively. While the not significant interaction shows that none of the hemifield-condition combinations was significantly different from the others, the significance in the factor condition shows that ctcOwnB indeed yielded a significantly better performance than ctcOther—when separately analyzed for the two hemifields. The 50% larger QE in ctcOther for the rear targets further supports this evidence.

A similar situation was revealed by the RM ANOVA performed on the PEs with factors target hemifield (frontal targets: -30° to 90°, rear targets: 90° to 210°), condition (ctcOwnB and ctcOther), and their interaction. The interaction ($p = 0.012$) was significant and the post-hoc test showed that while for the frontal targets, the difference between the conditions was not significant (32.4° for ctcOther and 33.6° for ctcOwnB), for the rear targets, significantly ($p < 0.01$) larger PE occurred in the ctcOther (40.8°) than in the ctcOwnB (33.7°) condition. Thus, the individualization of the CTC systems was able to substantially reduce the PE for rear targets.

All in all, the presented analysis demonstrates that in mismatched CTC systems, the sagittal plane localization performance improves when individualized CTC filters are considered, especially for targets placed behind the listener.

Channel Separation

The channel separation (CS) was calculated according to eq. (5.11) for all conditions and listeners. The CS, averaged in the frequency range 0.3 to 8 kHz (AKEROYD et al., 2007; PARODI and RUBAK, 2011) is shown in table 6.5 for each listener and condition. The last two rows of table 6.5 show the corresponding $\widehat{CS}$—calculated according to eq. (5.12) for the left ear and by analogy for the right ear—averaged over the same frequency range and over both ears.

For ctcOwn, the CS was large, on average 68.4 dB, and in the range of those reported previously for matched CTC systems (BAI and LEE, 2006; AKEROYD et al., 2007). For ctcOwnB, the CS was substantially lower, on average 14.7 dB, and in the range of $\widehat{CS}$ for both measured HRTF sets (14.9 and 14.5 dB). This indicates that individualized but mismatched CTC systems and reproduction systems without any CTC yield similar averaged CS.

For ctcOther, the CS was on average 15.0 dB and in the range of those reported previously for nonindividualized CTC systems (AKEROYD

Frequency Range CS	0.3 to 8 kHz									0.3 to 2 kHz	4 to 16 kHz
	NH12	NH14	NH15	NH57	NH62	NH64	NH68	NH72	Average ± Std. Dev.	Average ± Std. Dev.	Average ± Std. Dev.
ctcOwn	70.9	67.6	67.9	67.9	68.3	67.0	68.8	68.6	68.4 ± 1.2	50.4 ± 2.2	58.5 ± 2.6
ctcOwnB	16.2	13.5	11.5	15.6	19.2	18.2	11.4	12.3	14.7 ± 3.0	16.5 ± 3.4	14.2 ± 1.2
ctcKemar	17.2	15.6	16.8	15.9	18.1	18.0	12.6	18.9	16.6 ± 2.0	16.5 ± 1.9	12.6 ± 0.6
ctcNH57	13.6	17.4	17.9	-	13.9	16.3	11.9	13.7	14.9 ± 2.3	15.2 ± 2.8	13.2 ± 0.6
ctcNH64	17.2	15.1	16.1	16.2	17.7	-	11.0	17.1	15.8 ± 2.3	16.6 ± 2.3	13.6 ± 1.5
ctcNH68	14.5	11.5	12.0	11.9	12.2	11.0	-	12.9	12.3 ± 1.2	12.4 ± 2.1	11.9 ± 0.7
ctcNH12	-	13.6	13.9	13.6	16.3	17.2	14.5	18.9	15.4 ± 2.1	14.1 ± 1.2	14.5 ± 1.1
binOwn	16.4	15.3	15.1	14.4	14.3	14.5	13.8	14.6	14.8 ± 0.8	8.1 ± 0.3	17.1 ± 1.1
binOwnB	15.3	14.6	14.1	15.3	14.6	13.9	13.8	14.8	14.5 ± 0.6	8.2 ± 0.3	17.4 ± 1.4

Table 6.5: Channel separation CS in dB averaged over three frequency ranges. The last two rows show the natural channel separation $\overline{CS}$ averaged over both ears. Conditions not tested in the localization experiments are shown italic.

et al., 2007, average of 17.1 dB). It was also similar to that for ctcOwnB (14.7 dB), which might lead to the conclusion, that an individualization of the CTC systems is not necessary at all. Note that such a conclusion would not be consistent with the results from the previous section.

One reason for the similar CS in the individualized and nonindividu-alized CTC systems might be the choice of the frequency range used for averaging the frequency-dependent CS. In order to investigate the use of other frequency ranges, the CS was averaged over frequencies from 0.3 to 2 kHz (low-frequency CS) and from 4 to 16 kHz (high-frequency CS). Both low- and high-frequency CSs averaged over the listeners are shown in the right-most column of table 6.5. As averages over listeners, low- and high-frequency CSs showed a similar trend to the mid-frequency CS (0.3 to 8 kHz) when compared across the conditions. The correlation coefficient between all low-frequency and high-frequency CSs was 0.98.

Such a high correlation might arise, however, because of large and small CSs in the matched and mismatched conditions, respectively, and such large differences may dominate the correlation. Thus, the correlation was calculated separately for the matched and mismatched conditions. For the matched conditions only, the correlation coefficient between the low-frequency and the high-frequency CSs was 0.51. For the mismatched conditions only, the correlation coefficient was 0.24. A further comparison of the CS and $\widehat{CS}$ revealed that for the low-frequency range, the average CS was 15.0 dB and thus, larger than the corresponding average $\widehat{CS}$ of 8.15 dB. This indicates that the CTC indeed increased the CS in the frequencies below 2 kHz. For the high-frequency range, the average CS was 13.2 dB and thus *smaller* than the corresponding average $\widehat{CS}$ of 17.25 dB. This means that the CTC actually *decreased* the CS in the frequencies above 4 kHz for the tested CTC setup. Thus, if no CTC is applied in the frequency range above 4 kHz, the mismatched CTC systems tested in this experiment would show a larger CS.

This finding is in agreement with GARDNER (1997), who limited his CTC system to the low frequencies only. This observation can also explain the results from BAI et al. (2007), who band-limited their CTC to 6 kHz and obtained a lateral localization performance similar to that of the full-bandwidth CTC. While their choice for the band limitation was based on computational issues, the lack of the mismatched CTC at higher frequencies, and thus, no decrease in the CS at these frequencies might also have contributed to their findings. Generally, it seems that a frequency-dependent amount of CTC might be useful in order to avoid a decrease of the CS in mismatched CTC systems.

Localization Performance: Channel Separation

On the one hand, one quality aspect of a CTC system is the localization performance the system is able to provide. On the other hand, the CS is usually employed to describe the general quality of a CTC system. However, not much is known about the relation between the CS and localization performance.

The CS is calculated between the two ears and is, in principle, an interaural metric. Therefore, it might indeed have the potential to describe the horizontal plane localization performance, which also depends on interaural cues. The ITDs, being the most salient cues for sound localization in the horizontal planes (WIGHTMAN and KISTLER, 1992) are assumed to contribute in frequencies up to approximately 2 kHz (MACPHERSON and MIDDLEBROOKS, 2002). The contribution of ITDs to the localization performance was evaluated by comparing the horizontal plane performance with the low-frequency CS (0.3 to 2 kHz). The ILDs, also salient cues for the horizontal plane localization, are large in frequencies above approximately 3 kHz. The contribution of both ILDs and ITDs was evaluated by means of the mid-frequency CS (0.3 to 8 kHz). Even though CS is an interaural metric, the use of the CS to describe the sagittal plane localization performance was also investigated. The spectral cues, being the most salient cues for the sagittal plane localization (LANGENDIJK and BRONKHORST, 2002; MACPHERSON and MIDDLEBROOKS, 2002) are assumed to contribute the most in the frequency range from 4 to 16 kHz (CARLILE and PRA-LONG, 1994; PERRETT and NOBLE, 1997; MIDDLEBROOKS, 1999b). Thus, the sagittal plane localization performance was compared with the high-frequency CS (4 to 16 kHz).

At first sight, the relation between the CS and the performance seems to be weak. For example, for ctcOwn, ctcOwnB, and ctcOther, the average QE was 8.0%, 13.6%, and 16.9%, respectively, correspondent to a mid-frequency CS of 68.4, 14.7, and 15.0 dB, respectively. While from ctcOwn to ctcOwnB, the increase in QE is well represented by the decrease in CS, the further increase in QE from ctcOwnB to ctcOther is not. Generally, the CS in the range of 50 dB corresponds to a good localization performance. However, smaller CS (in the range between 13 to 18 dB) did not provide any statement on the localization performance. One example is NH72, who for quite different CSs (12.3 and 17.1 dB) showed nearly the same QE (23.2% and 23.9%). Other example is NH15, who for similar CSs (11.5 and 12.0 dB) showed completely different QEs (5% and 27.1%). Note that NH15 also showed a QE of 6.8% for the matched condition with a CS of 67.9 dB. This demonstrates the rather complex relation between CS and the localization performance in terms of QEs.

In order to estimate the statistical relation between the CS and the localization performance, the correlation between the CS and the localization errors is analyzed. The correlation coefficients calculated for the mid-, low, and high-frequency CS, are shown in table 6.6. For all tested conditions, the correlation coefficient between mid-frequency CS and QE, PE, and LE was -0.35, -0.32, -0.43, respectively (all significant, p < 0.025). Similar correlations resulted for the low- and high-frequency CS. Such weak correlations suggest that the CS is generally not a good predictor for the localization performance.

The low correlations between the CS and localization performance could, however, be put down to the listener-individual performance in the localization task. In order to compensate for the individual performance, the correlation coefficients were calculated between the CS and the performance relative to that obtained from the binOwn condition. All correlation coefficients increased (see table 6.6), with the largest coefficient being at -0.49. Hence, CS seems to be a poor predictor for the localization performance even when compensated for the listener-individual localization performance.

Since the localization performances of the matched and mismatched CTC systems differ tremendously, the CS might better correlate with the performance when compared separately for the matched and mismatched CTC conditions. For the matched condition (ctcOwn in table 6.6), the most correlation coefficients were low and not significantly different from zero, i.e. were uncorrelated. The only significant (p = 0.018) correlation coefficient (-0.79) was found between the high-frequency CS and the LE relative to that obtained for binOwn. This might suggest that in matched CTC systems, the high-frequency CS is able to predict the horizontal plane localization performance relative to the listener-individual performance.[10]

For the mismatched CTC systems, the largest significant correlation coefficients were -0.50 (QE and mid-frequency CS), -0.33 (PE and low-frequency CS), and -0.60 (LE and mid-frequency CS). These correlations did not improve when the relative localization performance was considered and show the extent to which CS might act as a predictor for the localization performance. Especially for the horizontal plane localization

---

[10]Note that despite the statistical support, this correlation is based on a small sample size (n=8) and that such a conclusion is to be treated with caution.

Frequency Range	0.3 to 8 kHz (mid)			0.3 to 2 kHz (low)			4 to 16 kHz (high)		
Condition	QE	PE	LE	QE	PE	LE	QE	PE	LE
all tested	**-0.35**	**-0.32**	**-0.43**	**-0.35**	**-0.33**	**-0.43**	**-0.32**	**-0.31**	**-0.38**
all tested. relative to bi-nOwn	**-0.37**	**-0.36**	**-0.45**	**-0.39**	**-0.38**	**-0.49**	**-0.35**	**-0.35**	**-0.42**
matched	0.00	-0.34	-0.31	0.21	0.03	0.51	0.30	-0.40	0.26
matched. relative to bi-nOwn	-0.25	-0.35	-0.27	-0.57	-0.36	-0.43	-0.25	-0.33	**-0.79**
mismatched	**-0.5**	-0.28	**-0.6**	**-0.35**	**-0.33**	**-0.42**	**-0.33**	-0.16	-0.27
mismatched. relative to bi-nOwn	**-0.48**	-0.20	**-0.55**	**-0.38**	-0.24	**-0.53**	-0.33	-0.02	**-0.25**

Table 6.6: Correlation coefficients for the correlation between the localization errors and the channel separation. Coefficients significantly ($p < 0.05$) different from zero are shown bold. The matched condition was ctcOwn. The mismatched conditions were ctcOwnB and ctcOthers.

performance, the correlation of -0.6 might be useful in further evaluations of CTC systems, depending on the application criteria. For the sagittal plane localization performance, the CS seems to be a poor predictor. This is not surprising, considering that monaural, not interaural, cues are the most salient cues for the sagittal plane localization.

6.2.3 Discussion

In experiment II the sound-localization performance in CTC reproduction systems was studied and its CTC filters were calculated under various conditions. The performance was compared to the baseline binaural condition. Channel separation, an objective measure for the quality of a CTC system, was calculated for the tested conditions and compared with the localization performance.

Under binaural conditions, the localization performance in terms of quadrant errors, polar errors, and lateral errors was within the range of previously reported performance. This was also the case when they were tested using HRTFs obtained from a measurement that was repeated approximately five years later, even though training was conducted only with the first HRTF set. This suggests that the human auditory localization system is robust to HRTF measurement variability—at least for the measurement setup used in this experiment.

With the matched CTC systems, the performance was similar to that from the binaural conditions. With the individualized but mismatched CTC systems, where CTC filters were based on the repeated HRTF measurements, the listeners showed a degraded localization performance in terms of larger lateral, polar, and quadrant errors. This shows that the propagation paths from the loudspeakers to the ears must *exactly* match the filters in a CTC system in order to provide localization performance at a similar level as the binaural reproduction. The direction-dependent analysis of the localization performance showed that in the mismatched CTC systems, the performance deteriorated especially for targets placed outside the loudspeaker span and/or behind the listener. With the nonindividualized CTC systems, the quadrant errors further increased for the rear targets and the performance for the frontal targets was in the range of that for the individualized but mismatched CTC system.

These findings show that for targets placed within the loudspeaker span and in the same hemisphere as the loudspeakers, the quality of the CTC system is not critical regarding localization and the amount of CTC can be reduced in order to provide a better timbre reproduction.

This might lead to a deteriorated apparent source width, though. Much attention, however, should be attached to the CTC systems for targets placed at directions outside the loudspeaker span. In particular for the rear targets, a work around to the currently unachievable matched CTC system, could be a second CTC system with loudspeakers placed behind the listener (PARODI and RUBAK, 2011). This appears indeed to be a promising approach. For the more lateral targets, additional loudspeakers at lateral positions might help. They could, combined with the loudspeakers in the rear, form a ring of loudspeakers around the listener. Such a system would have to consider all available loudspeaker combinations[11] to choose the most adequate CTC filter for each source to be reproduced and might thus be seen as an extension of the vector base amplitude panning (VBAP, PULKKI, 1997) to binaural reproduction. To the best of the author's knowledge, such a combination of systems has not been scientifically investigated yet.

A common quality metric for CTC systems is the channel separation. The results from this experiment show a substantial difference in channel separation between the matched and the mismatched CTC systems. However, channel separation was similar in both individualized and nonindividualized mismatched CTC systems, even though the sagittal plane localization performance was not. For the mismatched CTC systems, the channel separation was in the range of the natural channel separation provided by a stereophonic reproduction. The mismatched CTC systems improved the channel separation in frequency range below approximately 2 kHz but degraded the channel separation in the frequency range above approximately 4 kHz, suggesting that mismatched CTC should be avoided in the higher frequency regions. The matching had only little impact on the low-frequency channel separation. Hence, future efforts with regard to the matching should focus on the mid- and high-frequency regions, at least for the tested loudspeaker span.

A generally weak correlation (up to -0.35) was observed between the channel separation and the sagittal plane localization performance. This was also the case (up to -0.39) when compensating for the listener-individual localization performance in the binaural condition. The correlation increased to -0.5 when only mismatched CTC system were considered. This confirms the evidence that channel separation, being an interaural metric, is not an appropriate predictor for the sagittal plane localization performance.

For the horizontal plane, a better correlation between channel separation and localization performance could be expected. It was -0.49 in

---

[11] A method for the dynamic implementation of such filters is described in section 5.3.

general and increased to -0.79 when only matched CTC systems were considered. This correlation was between the high-frequency channel separation and the lateral errors relative to the baseline performance, it was significantly correlated, however, it was based on only eight samples. For mismatched CTC systems, the correlation of -0.6 was found for a more convincing sample size of 40 samples. Such a correlation indicates that the channel separation might be indeed useful when evaluating mismatched CTC systems with respect to the horizontal plane localization.

7

Conclusion

7.1 Summary

The main objective of this work was to improve the quality of binaural-based virtual acoustics systems. One of the key aspects to achieve this goal is individualization. Therefore, two components of individualized binaural technology were addressed in this thesis: the efficient acquisition of individual head-related transfer functions (HRTFs) and the adequate individual equalization of binaural reproduction systems.

This work started with the design of a measurement setup for the fast acquisition of individual HRTFs. The proposed solution was to construct a setup composed of an arc that can hold up to 40 loudspeakers and a turntable to rotate the subject inside that arc. This combination is assumed to be the best trade-off between hardware costs and measurement duration.

Binaural-based virtual reality systems commonly neglect the effect of near-field HRTFs. This is, nevertheless, a very important aspect when it comes to improving the realism of acoustic scenes. The range extrapolation technique, based on acoustic spherical holography, makes it possible to describe the HRTF's distance dependence based on measurements at a single distance. This setup is one of the first of its kind designed to be compatible with the range extrapolation technique. To avoid reflections coming from the supporting arc, it was built as a truss structure, expected to be acoustically transparent at the audible frequency range. The loudspeakers were designed in a drop-like shape to avoid edge diffraction. Finally, care was taken to choose a loudspeaker driver with broadband response that radiated as similar as possible to an omnidirectional source.

A reduced measurement time is not only more comfortable for the subject being measured, it can also help to reduce the measurement variability. The use of multiple loudspeakers and excitation signals in parallel will already produce a shorter total measurement time. Instead of orthogonal pseudo-random sequences, which are very sensitive to nonlinearity, the more robust multiple exponential sweep method (MESM) was used. This method was then further optimized, taking into consideration the time structure of the measured impulse response and allowing a more flexible distribution of the unwanted harmonic impulse responses along the raw impulse response. Measuring an HRTF dataset with 40 positions in elevation and 100 positions in azimuth in a sequential manner with an exponential sweep of length 1.34 s (assuming the turn table takes 0.3 s to reach its next position), would take 90 min. The same HRTF dataset measured with the original MESM would take just over 12 min to complete and applying the optimizations described in this work would reduce the measurement time even further, to less than 6 min. The use of interleaved sweeps has no influence on the obtained signal-to-noise ratio.

Measurements made with the MESM will result in a raw interleaved impulse response that needs further post-processing. After extracting each direction's transfer function, the signals must be further equalized to eliminate the influence of the transducers. At this point, the range extrapolation can be applied to account for the radial dependency of the HRTF. To verify the quality of the newly developed setup, an artificial head, whose HRTF dataset had already been measured with a previous generation setup composed of only one loudspeaker attached to a moving arm, was also measured with the newly constructed setup. The comparison of time, frequency, and spatial data showed a good agreement between the two measurement setups. The measurement with the new setup took indeed less than 6 min to complete, making this setup the fastest existing individual HRTF measurement system the author is aware of.

Individual HRTFs provide the basis for the binaural synthesis. After the binaural signals have been synthesized using these HRTFs, they must now be adequately played back to the listener. Binaural signals can then be reproduced either via headphones or via loudspeakers.

Reproduction via headphones is the more straightforward of the two methods, as headphones are able to feed each binaural signal directly to the respective ear. However, to deliver an authentic auditory impression without additional spectral coloration, the reproduction via headphones must be adequately equalized. Repeated measurements of the headphone transfer function (HpTF) confirmed that when the listeners are allowed to place the headphones themselves, at what they consider to be the most comfortable position, then HpTF variance drops considerably. On the other hand, measurements with several listeners also confirmed that HpTF varies considerably among subjects. A framework was developed for the design of individual headphone equalization filters. It includes several measurements (about ten) of the listener's HpTF, where the listeners are asked to take off the headphones in between each measurement. The magnitude spectra of these HpTFs are averaged and deep spectral dips are smoothed. The resulting magnitude spectrum is then inverted and finally the equivalent minimum-phase spectrum is obtained through the Hilbert transform.

The quality of the proposed equalization filters was verified with the aid of perceptual tests. The first part of this experiment consisted of a direct comparison (in a three-alternative forced-choice setup) of a stimulus reproduced through a loudspeaker—played in anechoic conditions—and its binaural synthesis reproduced via individually equalized open-type circumaural headphones. Three stimuli were used for this test: a pulsed pink noise, a speech sample and a music sample. Results showed that at least 50% of all listeners could not distinguish between the reproduction methods when hearing to speech (error rate: 38.17%) and music (error rate: 35.10%). Meanwhile, when listening to the pulsed pink noise—which was the only stimulus that contained spectral components above 10 kHz—listeners made significantly less mistakes (error rate: 16.74%), which suggests that stimuli with dominant high frequency components tend to allow an easier distinction between reproduction methods. This result is reasonable as HpTFs show higher variance exactly at this frequency range.

The second part of this experiment was an indirect comparison, where listeners were asked to say whether the presented stimulus—this time only the pulsed pink noise—came from the loudspeaker or the headphones, excited as in the first part of this experiment. Results showed that listeners were guessing at almost all times (error rate: 51.03%), indicating that the binaural signals reproduced via individually equalized headphones sounded natural and authentic. Thus, even though differences between the original anechoic sound source and its binaural auditory display could be heard when the stimulus contained high frequency components, these differences were not big enough to allow the listeners to state clearly whether the auditory event that was binaurally reproduced via individually equalized headphones was actually coming from the headphones and not from an external source.

The other way to play back binaural signals to a listener is via (at least two) loudspeakers. This method is commonly preferred for applications where the use of headphones might hinder the sense of immersion, e.g. in virtual reality environments. Binaural reproduction via loudspeakers suffers from crosstalk, which mixes the spatial cues contained in the binaural signal. This effect can be compensated for by using crosstalk cancellation (CTC) filters, which generate the desired channel separation between the listener's ears. In virtual reality applications, users should have complete freedom of movement. For such systems, a dynamic CTC with multiple loudspeakers is required and the CTC filters must be constantly updated according to the tracked head position. As frequency-domain calculations are usually more efficient (and faster) than their time-domain counterparts, dynamic CTC systems tend to be implemented in frequency-domain. This has the drawback that such filters display noncausal artifacts, which is not the case in time-domain calculations.

A framework was proposed for the calculation of causal CTC filters in the frequency-domain. An approximation of the causal solution is obtained by using a time-domain calculation which is in turn based on the Wiener-Hopf decomposition. It is known that CTC filters show a large gain boost at certain frequencies and regularization is used as a gain limiter. However, the regularization has the drawback that it adds noncausal artifacts both in the CTC filters and in the resulting impulse response in the ears. These noncausal artifacts can be eliminated—or rather be transformed in causal artifacts—through the minimum-phase regularization. Instead of the channel-dependent solution proposed with the original minimum-phase regularization method, the proposed framework applies a new global minimum-phase regularization strategy.

Dynamic CTC systems need multiple loudspeakers to be able to switch between active loudspeakers, thus avoiding instability of the CTC filters. It was shown that simple panning between two configurations can affect the system's resulting frequency response. The proposed framework incorporates spatial fading in the filter calculation stage through a weighted matrix inversion, which provides a smooth transition between active loudspeakers.

CTC filters are calculated from the transfer functions between loud-speakers and listener's ears, i.e., the HRTFs. Therefore, the CTC filters are subject to individualization as well. Nevertheless, many CTC systems use generic transfer functions for its filter calculation. A localization test was conducted to evaluate the influence of individualization on the localization performance of CTC systems. So far, only one similar evaluation has been conducted, but it used an auditory model simulation instead of a perceptual test. The localization performance was tested with regard to quadrant errors (QE), polar errors (PE), and lateral errors (LE). Listeners had their HRTF measured twice (within an interval of five ears). Acoustic targets were Gaussian white noises, filtered with the listener-specific directional transfer functions. Baseline tests showed that the localization performance was within the range of previously reported performance for both HRTF sets (set 1: LE=10.7°, PE=31.0°, QE=8.6%; set 2: LE=10.4°, PE=31.6°, QE=7.1%), even though training was provided using only the first of these sets, suggesting that the human auditory localization system is robust to small HRTF measurement variability.

In this experiment, CTC systems were virtually rendered and presented via headphones. Results showed that individualized matched CTC systems (the same HRTFs are used for the filter calculation and the loudspeaker rendering) provided performance similar to that from the binaural listening (LE=10.3°, PE=31.8°, QE=8.0%). For individualized mismatched systems (two different HRTF datasets from the same listener are used for the filter calculation and the loudspeaker rendering) the localization performance deteriorated (LE=13.0°, PE=34.2°, QE=13.6%). And for nonindividualized mismatched systems (the CTC filters are calculated with the HRTFs from other listeners) the sagittal localization errors increased further (PE=36.7°, QE=16.9%). The direction-dependent analysis showed that mismatch and lack of individualization yielded a degraded performance for targets placed outside of the loudspeaker span (LE=17.2°) and behind the listeners (PE=40.8°, QE=26.1%), indicating the relevance of individualized CTC systems for such targets.

It is commonly assumed that binaural reproduction through loudspeakers will only work if the CTC filters can provide a sufficient channel separation. Thus, channel separation is very often used as a predictor for the quality of a CTC system. The channel separation was calculated for all conditions evaluated in this localization test for different frequency ranges. The results showed a substantial difference in channel separation between the matched and the mismatched CTC systems, but similar values in both individualized and nonindividualized mismatched CTC systems, which does not match the observed variations in localization performance. It was observed that the mismatched CTC systems improved the channel separation in frequency range below approximately 2 kHz, but degraded the channel separation in the frequency range above 4 kHz, an observation that is in agreement with the practical knowledge that mismatched CTC should be avoided at the higher frequency regions. Results showed that channel separation might be indeed useful for the evaluation of mismatched CTC systems with respect to the horizontal plane localization, but it is not an appropriate predictor for the sagittal plane localization performance.

All in all, this thesis extended the current knowledge on the efficient measurement of individual head-related transfer functions and highlighted the importance of individual equalization filters in binaural reproduction, both via loudspeakers and headphones. Moreover, an integrated framework for the calculation of such equalization filters was presented.

7.2 Outlook

Some aspects in the field of individual binaural technology could be improved in the course of this thesis, but a number of other questions were left unanswered and many new questions were raised. This section presents some ideas that could further improve the quality of the acquisition and post-processing of individual HRTFs as well as the quality of binaural reproduction methods.

The proposed individual HRTF measurement setup used a head-rest to stabilize the listener's head. This apparatus did reduce the listener's movement during measurement, but could not completely eliminate it. A possible solution to compensate for the influence of small head movements during measurement would be to track the position of the listener's head and compensate for the observed movements in a post-processing step, using a measurement grid for the spherical harmonic interpolation made with the tracked head position at the time of each measurement.

The chosen design of the arc proved not to be ideal. From a mechanical point of view, the used truss structure was too delicate and the solder joints constantly gave away. From an acoustical point of view, the arc itself vibrated during measurement, acting as an unwanted acoustic source. Even though the designed drop-like loudspeakers did have an adequate radiation pattern, they reflected the sound coming from the neighboring speakers. Thus, another design of the arc should be investigated, eventually attaching the loudspeaker to a more rigid arc with a continuous acoustically absorbing material. Furthermore, the loudspeaker drivers displayed a time-variant behavior, unacceptable for this kind of measurement. A new loudspeaker driver should be chosen that does not display this kind of behavior.

The truncation of the spherical harmonic order during interpolation or range extrapolation results in a spatial smoothing of the data. It is known that HRTFs smoothed in frequency-domain are still able to provide correct spatial impression (KULKARNI and COLBURN, 1998;

XIE and ZHANG, 2010), so it is also possible that a spatially smoothed HRTF dataset will also provide a correct spatial impression. Thus, a psychoacoustic evaluation of the required spherical harmonic order to reconstruct a perceptually correct HRTF dataset is necessary, as currently available studies on the required spherical harmonic order that adequately describe an HRTF dataset, take only physical aspects into account. It would also be important to investigate whether this limit varies with distance, as this would directly influence the truncation limits of the range extrapolation stage.

Even though the constructed HRTF measurement setup is already quite fast, it could be made even faster if the listeners were constantly rotated during the measurement as is the case in the continuous HRTF measurement methods proposed by ENZNER (2009) and the plenacoustic interpolation method proposed by AJDLER et al. (2007). Theoretical calculations suggest that a dynamic measurement scheme using the MESM would reduce the measurement duration of 4000 HRTFs from the current 6 min to a mere 2 min, keeping the robustness to nonlinearity. However, the HRTFs measured that way will be spatially blurred. It is still to be evaluated if this blur is of perceptual relevance. Nevertheless, first steps towards compensating the influence of rotation in the dynamic measurement have been reported by KRECHEL (2012).

The fact that loudspeakers adjacent in elevation are sitting on opposite halves of the arc can lead to strong phase ripples in the HRTF's spatial representation, which are harmful to the spherical harmonic processing of the data. This will happen if the subject being measured is not placed with its longitudinal axis exactly centered with the turntable's rotation axis. As such a precise positioning of the listener is practically impossible, other methods should be evaluated on how to correct this effect in a post-processing stage. One possibility would be to conduct a dynamic reference measurement, as proposed by KRECHEL (2012). Another alternative, possibly more precise, would be to search for the acoustic center of the points from each half of the arc independently—e.g. with a search algorithm as the one proposed by ZIEGELWANGER (2012)—and then shift both halves together.

The limited number of measurement points can lead to spatial aliasing when post-processing the HRTF dataset. A possible way to deal with this spatial aliasing would be to use the compressive sensing framework. A preliminary investigation showed that spherical harmonics are not an adequate basis to describe the HRTFs in a sparse manner (MASIERO and POLLOW, 2010). Work has still to be done in searching for an adequate basis for the sparse representation of HRTFs. A possible basis could be spherical wavelets or the Slepian functions.

Besides individual HRTFs, the binaural reproduction could also be improved. It is clear that binaural reproduction via headphones is a well-established technique and this thesis even confirmed that auditory displays presented via individually equalized headphones sound realistic and authentic. However, it would be interesting to evaluate how the lack of equalization will actually influence the auditory impression as, e.g., the influence of headphone equalization in localization is still debatable. This research could lead to the design of a new pair of headphones, which could provide an authentic auditory impression without individual equalization, thus facilitating the introduction of binaural technology in the consumer market.

On the other hand, the binaural reproduction via loudspeakers still has some hurdles to clear before it is ready for the consumer market. The first, and probably easiest of them, is an efficient low cost head tracking device. This could be easily implemented with a camera and face tracking software. First steps in this direction have been taken by FUNDALEWICZ (2012).

A second very important aspect—deliberately ignored throughout this thesis—is the influence of the reproduction room, as unwanted reflections play a major role in the quality of the reproduced binaural signal. An interesting approach proposed to control this effect is to generate CTC filters that also compensate for room reflections, but only on a time range where reflections are above the temporal masking threshold of the human auditory system (JUNGMANN et al., 2012). The use of psychoacoustical knowledge to lessen the restrictions imposed on CTC systems seems to be a promising field of studies.

In this context, virtual reality reproduction systems could profit from a hybrid scheme, ideally combining the advantages of different spatial audio reproduction techniques. Such a hybrid method was first proposed by PELZER; MASIERO, and VORLÄNDER (2011), who suggested the use of a binaural CTC system to reproduce only the direct sound and early reflections and the use of a system like low-order ambisonics to reproduce the reverberant tail of a simulated room impulse response.

A further possible improvement of CTC systems can be derived from the results presented in this thesis, namely that localization of sources within the loudspeakers' span was not significantly deteriorated by a mismatched CTC system. First, the speculation that this situation occurs because central targets in a mismatched CTC system are localized with the same psychoacoustical process used to localize phantom sources in a stereophonic reproduction system should be investigated. Furter, in a setup with multiple loudspeakers, each reflection of a room impulse response could be played back by the group of loudspeakers chosen in accordance with the direction this reflection is arriving from, much in the same way as vector base amplitude panning does its active triangle selection.

Finally, as it was shown that channel separation does not seem to be an adequate predictor for the localization performance provided by a CTC system, a new type of CTC quality predictor should be developed. Auditory models could come in handy to help define such a predictor.

A

Regularization as a Gain Limiter

FARINA (2007) suggests the use of a time-packing filter to control the size of an inverse filter's impulse response. He achieved this by conducting a regularized inversion in frequency-domain, as follows

$$C(z) = H(z)^* / (H(z)^* H(z) + \mu).$$ (A.1)

Regularization acts, however, not only as a time-packing tool. It also works as a gain limiter in the frequency-domain. This can be verified by taking the derivative of the amplitude of $C(z)$ in relation to the amplitude of $H(z)$

$$\frac{\partial |C(z)|}{\partial |H(z)|} = \frac{-|H(z)|^2 + \mu}{(|H(z)|^2 + \mu)^2}.$$ (A.2)

From eq. (A.2), when $|H(z)|^2 = \mu$, the maximum value of $|C(z)|$ is

$$|C(z)| = \frac{1}{(2\sqrt{\mu})}.$$ (A.3)

Thus, if $|C(z)|$ should be no greater than x dB, μ should be chosen to be

$$\mu = \frac{1}{(2 \cdot 10^{x/20})^2}.$$ (A.4)

Also in the multi-channel regularized inversion μ can be understood as a gain limiter. We know that the Euclidean norm of a matrix is given by

$$\|\boldsymbol{H}\|_2 = \sigma_{\max}(\boldsymbol{H}),$$ (A.5)

where $\sigma_{\max}$ is the largest singular value of $\boldsymbol{H}$. Assuming that the singular value decomposition from $\boldsymbol{H}$ is

$$\boldsymbol{H} = \boldsymbol{U\Sigma V}^*,$$ (A.6)

where $\boldsymbol{U}$ and $\boldsymbol{V}$ are unitary matrices and $\boldsymbol{\Sigma}$ is a diagonal matrix containing the singular values of $\boldsymbol{H}$, then the regularized matrix inversion

$$C = (\boldsymbol{H}^*\boldsymbol{H} + \mu\boldsymbol{I})^{-1}\boldsymbol{H}^*.$$ (A.7)

can be rewritten as

$$C = V(\Sigma'\Sigma + \mu I)^{-1}\Sigma'U^*, \qquad (A.8)$$

where Σ' is the transpose of Σ and the singular values from C are thus

$$\sigma_i(C) = \frac{\sigma_i(H)}{(\sigma_i(H)^2 + \mu)}. \qquad (A.9)$$

GOLUB and VAN LOAN (1996) argue that the amplitude of the largest element of a matrix is smaller or equal to the Euclidean norm of this matrix, thus

$$\max_{i,j}|c_{i,j}(z)| \leq \|C\|_2 = \sigma_{\max}(C) = \frac{\sigma_j(H)}{(\sigma_j(H)^2 + \mu)}, \qquad (A.10)$$

where $\sigma_j(H)$ is the singular value of H that results in the largest singular value of C.

Taking the maximum of $\sigma_{\max}(C)$ with regards to $\sigma_j(H)$, as was done for the one channel case, results in

$$\max_{i,j}|c_{i,j}(z)| \leq \frac{1}{(2\sqrt{\mu})}. \qquad (A.11)$$

Thus, just as for the single channel inversion, also for the multi-channel inversion the regularization parameter μ acts as an upper bound for every resulting inverse filters.

B

Least-Square Minimization

If the transfer matrix H is underdetermined, i.e. it has more columns than rows, there will be an infinite number of CTC filter combinations that can drive the error energy d to zero. In this case, besides the minimization of the error energy, the *control effort*, i.e. the energy of the loudspeaker signals, is also minimized. This extra constraint added to the cost function leads to a single optimal solution to this minimization problem.

Such minimization requirements can be cast as a constrained optimization problem using Lagrange multipliers (NELSON and EL-LIOTT, 1995). The cost function to be minimized is now

$$J(z) = -v^* v - d^* \lambda - \lambda^* d, \tag{B.1}$$

where λ is a vector of Lagrange multipliers. This equation can be expanded to reveal its dependency on C.

$$\begin{aligned} J(z) &= -b^* C^* C b - b^* \left(H C - I \cdot e^{-z\Delta} \right)^* \lambda - \\ &\quad \lambda^* \left(H C - I \cdot e^{-z\Delta} \right) b. \end{aligned} \tag{B.2}$$

The filters for each ear are optimized independently. As the optimization depends on the input signal, KIRKEBY and NELSON (1999) suggest to set $b_j(n) = \delta(n)$, the Dirac delta function, as this gives the worst-case scenario for the optimization. The new cost function to be minimized is now

$$J(z) = -c_i^* c_i - \left(H c_i - y_i \right)^* \lambda - \lambda^* \left(H c_i - y_i \right), \tag{B.3}$$

where c_i is the ith column of C and y_i is the ith column of $I \cdot e^{-z\Delta}$.

The derivative with respect to a vector is defined as

$$\frac{\partial z}{\partial x} = \begin{bmatrix} \frac{\partial z}{\partial x_1} \\ \vdots \\ \frac{\partial z}{\partial x_n} \end{bmatrix}. \tag{B.4}$$

According to NELSON and ELLIOTT (1995), deriving J with respect to both c and λ results in

$$\partial J/\partial c = -2c_i - 2H^*\lambda, \tag{B.5}$$

$$\partial J/\partial \lambda = -2\left(Hc_i - y_i\right). \tag{B.6}$$

The minimum value of c_i is given by setting both derivatives to zero and substituting them into each other to isolate c_i. Assuming that HH^* is not singular, then

$$c_i = H^*\left(HH^*\right)^{-1}y_j, \tag{B.7}$$

that is equivalent in matrix form to

$$C = H^*\left(HH^*\right)^{-1}e^{-z\Delta}. \tag{B.8}$$

B.1 Regularized Least-Square Minimization

The Lagrange multipliers λ can be related to the effort made by the filters to satisfy the constraints imposed in eq. (B.3). This minimization can be regularize by restricting the effort made by the CTC filters, resulting in the new cost function

$$J(z) = -c_i^*c_i - \left(Hc_i - y_i\right)^*\lambda - \lambda^*\left(Hc_i - y_i\right) + \mu\lambda^*\lambda. \tag{B.9}$$

Equation (B.5) remains unaltered while eq. (B.6) changes to

$$\partial J/\partial \lambda = -2\left(Hc_i - y_i\right) + 2\mu\lambda, \tag{B.10}$$

which yields

$$C = H^*\left(HH^* + \mu I\right)^{-1}e^{-z\Delta}. \tag{B.11}$$

B.2 Weighted Regularized Least-Square Minimization

Substituting the ℓ_2 norm of c_i by the weighted norm eq. (5.32) in eq. (B.9) yields

$$J(z) = -c_i^*Zc_i - \left(Hc_i - y_i\right)^*\lambda - \lambda^*\left(Hc_i - y_i\right) + \mu\lambda^*\lambda. \tag{B.12}$$

The larger the weight z_i given for a given loudspeaker, the higher the effort made by the algorithm to minimize this loudspeaker's energy and thus the smallest the energy of the filters related to this loudspeaker.

Equation (B.10) remains unaltered while eq. (B.5) changes to

$$\partial J/\partial c = -2Zc_i - 2H^*\lambda, \tag{B.13}$$

which yields

$$C = Z^{-1}H^* \left(HZ^{-1}H^* + \mu I\right)^{-1} e^{-z\Delta} \tag{B.14}$$

as long as Z^{-1} exists, which is the case if Z is a diagonal matrix and $\forall z_i > 0$.

If, however, Z is not directly invertible, another solution, based in the method described in (RUFFINI et al., 2002) can be used. First $\partial J/\partial\lambda$ is multiplied by H^* and added to $\partial J/\partial c$, giving

$$(Z + H^*H)c_i = H(\lambda + y + \mu\lambda). \tag{B.15}$$

The dependency on λ is eliminated by multiplying $\partial J/\partial c$ by H, isolating λ and substituting it into eq. (B.15), which, after some algebraic manipulations, results in

$$C = [H^*H + (I - P)Z + \mu PZ]^{-1} e^{-z\Delta}, \tag{B.16}$$

where $P = H^*(HH^*)^{-1}H$.

List of References

AJDLER, T.; SBAIZ, L., and VETTERLI, M. (Sept. 2007). "Dynamic measurement of room impulse responses using a moving microphone." In: *J. Acoust. Soc. Am.* 122.3, p. 1636 (cit. on pp. 65, 149).

AKEROYD, M. A.; CHAMBERS, J.; BULLOCK, D.; PALMER, A. R.; SUMMERFIELD, A. Q.; NELSON, P. A., and GATEHOUSE, S. (2007). "The binaural performance of a cross-talk cancellation system with matched or mismatched setup and playback acoustics". In: *J. Acoust. Soc. Am.* 121.2, pp. 1056–1069 (cit. on pp. 3, 83, 116, 117, 120, 123, 127, 129–132, 134).

ALGAZI, V. R.; DUDA, R.; THOMPSON, D., and AVENDANO, C. (2001). "The CIPIC HRTF database". In: *Applications of Signal Processing to Audio and Acoustics, 2001 IEEE Workshop on the*, pp. 99–102 (cit. on pp. 2, 21, 23, 28).

ARETZ, M. (2012). "Combined Wave And Ray Based Room Acoustic Simulations In Small Rooms". PhD. RWTH Aachen University (cit. on p. 52).

ATAL, B. S.; HILL, M., and SCHROEDER, M. R. (1966). *Apparent Sound Source Translator* (cit. on p. 3).

BAI, M. R. and LEE, C.-C. (2006). "Objective and subjective analysis of effects of listening angle on crosstalk cancellation in spatial sound reproduction". In: *J. Acoust. Soc. Am.* 120.4, p. 1976 (cit. on pp. 83, 116, 117, 134).

BAI, M. R.; SHIH, G.-Y., and LEE, C.-C. (2007). "Comparative study of audio spatializers for dual-loudspeaker mobile phones". In: *J. Acoust. Soc. Am.* 121.1, p. 298 (cit. on p. 136).

BAUCK, J. L. and COOPER, D. H. (1992). "Generalized Transaural Stereo". In: *93rd AES Convention*. San Francisco, USA (cit. on pp. 81, 82).

BAUCK, J. L. and COOPER, D. H. (1993). "On Transaural Stereo for Auralization". In: *95th AES Convention*. Vol. 3728. New York (cit. on p. 3).

BAUCK, J. L. and COOPER, D. H. (1996). "Generalized transaural stereo and applications". In: *J. Audio Eng. Soc.* 44.9, pp. 683–705 (cit. on p. 77).

BAUER, B. B. (1961). "Stereophonic Earphones and Binaural Loudspeakers". In: *J. Audio Eng. Soc.* 9.2, pp. 148–151 (cit. on p. 3).

BEGAULT, D. R.; WENZEL, E. M.; LEE, A. S., and ANDERSON, M. R. (Oct. 2000). "Direct comparison of the impact of head tracking, reverberation, and individualized head-related transfer functions on the spatial perception of a virtual speech source". In: *108th AES Convention.* Vol. 49. 10 (cit. on p. 14).

BLAUERT, J. (1969). "Sound localization in the median plane". In: *Acustica* 22, pp. 205–213 (cit. on p. 18).

BLAUERT, J. (1997). *Spatial hearing: the psychophysics of human sound localization.* MIT Press (cit. on pp. 1, 15, 17, 18, 45, 47, 83, 104, 114).

BORISH, J. and ANGELL, J. B. (1983). "An Efficient Algorithm for Measuring the Impulse Response using Pseudorandom Noise". In: *J. Audio Eng. Soc.* 31 (cit. on p. 11).

BOUCHARD, M.; NORCROSS, S. G., and SOULODRE, G. (2006). "Inverse Filtering Design Using a Minimal-Phase Target Function from Regularization". In: *121st AES Convention.* San Francisco, USA (cit. on pp. 9, 13).

BOYD, S. and VANDENBERGHE, L. (June 2004). *Convex Optimization Theory.* New York: Cambridge University Press, p. 730 (cit. on p. 86).

BREEBAART, J. and KOHLRAUSCH, A. (2001). "The perceptual (ir) relevance of HRTF magnitude and phase spectra". In: *110th AES Convention.* Amsterdam, The Netherlands: Audio Engineering Society; 1999 (cit. on p. 47).

BRONKHORST, A. W. (1995). "Localization of real and virtual sound sources". In: *J. Acoust. Soc. Am.* 98.5, p. 2542 (cit. on pp. 2, 21, 23, 101).

BRUNGART, D. S. and RABINOWITZ, W. M. (Oct. 1999). "Auditory localization of nearby sources. Head-related transfer functions." In: *J. Acoust. Soc. Am.* 106.3 Pt 1, pp. 1465–79 (cit. on pp. 2, 24, 52).

BUCKLEIN, R. (1981). "The audibility of frequency response irregularities". In: *J. Audio Eng. Soc.* 29.3, pp. 126–131 (cit. on p. 72).

CARLILE, S. and PRALONG, D (1994). "The location-dependent nature of perceptually salient features of the human head-related transfer functions". In: *J. Acoust. Soc. Am.* 95, pp. 3445–3459 (cit. on p. 137).

CRUZADO, C. G. M. (2002). "Influence of the Acoustic Impedance of the Headphone on Psychoacoustic Effects". M.Sc. RWTH Aachen University (cit. on p. 69).

DANIEL, J. (2003). "Spatial Sound Encoding Including Near Field Effect: Introducing Distance Coding Filters and a Viable, New Ambisonic Format". In: *23rd International Conference: Signal Processing in Audio Recording and Reproduction.* Copenhagen, Denmark (cit. on p. 1).

DELLEPIANE, M; PIETRONI, N; TSINGOS, N; ASSELOT, M, and SCOPIGNO, R (2008). "Reconstructing head models from photographs for individualized 3D-audio processing". In: *27th Computer Graphics Forum.* Vol. 27. 7. Trier, Germany, pp. 1719–1727 (cit. on p. 2).

DIETRICH, P.; MASIERO, B., and VORLÄNDER, M. (2012a). "On the Optimization of the Multiple Exponential Sweep Method". In: *J. Audio Eng. Soc.* Submitted (cit. on pp. 21, 33, 36, 40).

DIETRICH, P.; MASIERO, B.; POLLOW, M.; KRECHEL, B., and VORLÄNDER, M. (2012b). "Time Efficient Measurement Method for Individual HRTFs". In: *Fortschritte der Akustik – DAGA.* Darmstadt, Germany (cit. on pp. 38, 40).

DRISCOLL, J. and HEALY, D. (1994). "Computing Fourier transforms and convolutions on the 2-sphere". In: *Advances in Applied Mathematics* (cit. on p. 47).

DURAISWAMI, R.; ZOTKIN, D. N., and GUMEROV, N. (2004). "Interpolation and range extrapolation of HRTFs". In: *2004 IEEE International Conference on Acoustics, Speech, and Signal Processing.* Vol. 4, pp. iv–45–48 (cit. on pp. 2, 18, 47, 48, 50, 51).

ENZNER, G. (2009). "3D-continuous-azimuth acquisition of head-related impulse responses using multi-channel adaptive filtering". In: *Applications of Signal Processing to Audio and Acoustics, 2009. WASPAA'09. IEEE Workshop on,* pp. 325–328 (cit. on pp. 65, 149).

EVANS, M. J.; ANGUS, J. A. S., and TEW, A. I. (1998). "Analyzing head-related transfer function measurements using surface spherical harmonics". In: *J. Acoust. Soc. Am.* 104.4, pp. 2400–2411 (cit. on pp. 47, 48).

FARINA, A. (2007). "Advancements in impulse response measurements by sine sweeps". In: *122nd AES Convention*. Vol. 122. Vienna, Austria (cit. on pp. 12, 153).

FASTL, H. and ZWICKER, E. (2007). *Psychoacoustics: Facts and Models*. Springer, p. 462 (cit. on p. 90).

FAZI, F. M. (2010). "Sound Field Reproduction". PhD thesis. University of Southampton, p. 297 (cit. on p. 9).

FELS, J. (2008). "From children to adults: How binaural cues and ear canal impedances grow". Ph.D. RWTH Aachen University (cit. on pp. 2, 13).

FELS, J. and MASIERO, B. (June 2011). "Binaural reproduction technologies for studies on dichotic and selective binaural hearing : Headphone reproduction". In: *Proceeding of Forum Acusticum*. Aalborg, Denmark (cit. on p. 67).

FIELDER, L. D. (2003). "Analysis of traditional and reverberation-reducing methods of room equalization". In: *J. Audio Eng. Soc.* 51.1/2, pp. 3–26 (cit. on p. 90).

FREEDEN, W. and WINDHEUSER, U. (Jan. 1997). "Combined Spherical Harmonic and Wavelet Expansion—A Future Concept in Earth's Gravitational Determination". In: *Applied and Computational Harmonic Analysis* 4.1, pp. 1–37 (cit. on p. 66).

FREELAND, F. P.; BISCAINHO, L. W., and DINIZ, P. S. R. (2007). "HRTF interpolation through direct angular parameterization". In: *Proc. 2007 IEEE Intern. Symposium on Circuits and Systems* May, pp. 1823–1826 (cit. on p. 47).

FUKUDOME, K; SUETSUGU, T; UESHIN, T; IDEGAMI, R, and TAKEYA, K (Aug. 2007). "The fast measurement of head related impulse responses for all azimuthal directions using the continuous measurement method with a servo-swiveled chair". In: *Applied Acoustics* 68.8, pp. 864–884 (cit. on p. 65).

FUNDALEWICZ, J. (2012). "Mobiles transaurales Wiedergabesystem mit videobasiertem Head-Tracking". Diploma. RWTH Aachen (cit. on p. 150).

GARDNER, W. G. (1997). "3-D audio using loudspeakers". PhD. Massachusetts Institute of Technology (cit. on pp. 3, 78, 83, 95, 116, 117, 136).

GARDNER, W. G. and MARTIN, K. D. (1995). "HRTF measurements of a KEMAR". In: *J. Acoust. Soc. Am.* 97.6, pp. 3907–3908 (cit. on p. 123).

GERZON, M. (1973). "PERIPHONY: WITH-HEIGHT SOUND RE-PRODUCTION". In: *J. Audio Eng. Soc.* 21.1, pp. 2–10 (cit. on p. 1).

GOLAY, M. (Apr. 1961). "Complementary series". In: *IEEE Transactions on Information Theory* 7.2, pp. 82–87 (cit. on p. 11).

GOLUB, G. H. and VAN LOAN, C. F. (1996). *Matrix Computations*. Johns Hopkins Studies in the Mathematical Sciences. Johns Hopkins University Press (cit. on p. 154).

GOUPELL, M.; MAJDAK, P., and LABACK, B. (Feb. 2010). "Median-plane sound localization as a function of the number of spectral channels using a channel vocoder". In: *J. Acoust. Soc. Am.* 127.2, pp. 990–1001 (cit. on p. 126).

GUILLON, P; ZOLFAGHARI, R; EPAIN, N; van SCHAIK, A; JIN, C. T.; HETHERINGTON, C; THORPE, J, and TEW, A (2012). "Creating the Sydney York Morphological and Acoustic Recordings of Ears Database". In: *2012 IEEE International Conference on Multimedia and Expo*, pp. 461–466 (cit. on p. 2).

HAMMERSHØI, D. and MØLLER, H. (2005). "Binaural Technique— Basic Methods for Recording, Synthesis and Reproduction". In: *Communication Acoustics*. Ed. by J. Blauert. Springer-Verlag, p. 379 (cit. on pp. 41, 46, 64, 71).

HARTMANN, W. M. (Nov. 1999). "How We Localize Sound". en. In: *Physics Today* 52.11, p. 24 (cit. on pp. 15, 19).

HOFMAN, P. M.; VAN RISWICK, J. G., and VAN OPSTAL, A. J. (Sept. 1998). "Relearning sound localization with new ears." In: *Nature neuroscience* 1.5, pp. 417–21 (cit. on p. 14).

JUNGMANN, J. O.; MAZUR, R.; KALLINGER, M.; MEI, T., and MERTINS, A. (Aug. 2012). "Combined Acoustic MIMO Channel Crosstalk Cancellation and Room Impulse Response Reshaping". In: *IEEE Transactions on Audio, Speech and Language Processing* 20.6, pp. 1829–1842 (cit. on pp. 97, 150).

KATZ, B. F. G. (Nov. 2001). "Boundary element method calculation of individual head-related transfer function. I. Rigid model calculation". en. In: *J. Acoust. Soc. Am.* 110.5, p. 2440 (cit. on p. 2).

KIM, S.-M. and WANG, S. (2003). "A Wiener filter approach to the binaural reproduction of stereo sound". In: *J. Acoust. Soc. Am.* 114.6, p. 3179 (cit. on p. 89).

KIM, Y.; DEILLE, O., and NELSON, P. A. (Oct. 2006). "Crosstalk cancellation in virtual acoustic imaging systems for multiple listeners". In: *Journal of Sound and Vibration* 297.1-2, pp. 251–266 (cit. on p. 82).

KIRKEBY, O. and NELSON, P. A. (1999). "Digital Filter Design for Inversion Problems in Sound Reproduction". In: *J. Audio Eng. Soc.* 47.7/8, pp. 583–595 (cit. on pp. 3, 81, 84, 90, 155).

KIRKEBY, O.; NELSON, P. A.; HAMADA, H., and ORDUNA-BUSTAMANTE, F. (Mar. 1998a). "Fast deconvolution of multichannel systems using regularization". In: *IEEE Transactions on Speech and Audio Processing* 6.2, pp. 189–194 (cit. on pp. 86, 89, 90).

KIRKEBY, O.; NELSON, P. A., and HAMADA, H. (1998b). "The 'stereo dipole' a virtual source imaging system using two closely spaced loudspeakers". In: *J. Audio Eng. Soc.* 46.5, pp. 387–395 (cit. on pp. 78, 86).

KISTLER, D. J. and WIGHTMAN, F. L. (1992). "A model of head-related transfer functions based on principal components analysis and minimum-phase reconstruction". In: *J. Acoust. Soc. Am.* 91.3, pp. 1637–1647 (cit. on pp. 47, 66).

KLEBER, J and VORLÄNDER, M (2001). "Messung von Gehöreingangsimpedanzen des freien Ohres und des abgeschlossenen Ohres mit Otoplastiken, Im-Ohr-Hörgeräten oder Kopfhörern". In: *Fortschritte der Akustik – DAGA* (cit. on p. 69).

KÖRING, J and SCHMITZ, A. (1993). "Simplifying Cancellation of Cross-Talk for Playback of Head-Related Recordings in a Two-Speaker System". In: *Acustica* 79.December 1992, pp. 221–232 (cit. on pp. 3, 93).

KRECHEL, B. (2012). "Schnelle Messung von individuellen HRTFs mit kontinuierlichen MIMO-Verfahren". M.Sc. RWTH Aachen (cit. on pp. 30, 31, 64, 65, 149).

KULKARNI, A. and COLBURN, H. (1998). "Role of spectral detail in sound-source localization". In: *Nature* 396.December, pp. 747–749 (cit. on pp. 47, 102, 148).

KUTTRUFF, H. (2000). *Room Acoustics*. CRC Press (cit. on p. 11).

LANGENDIJK, E. H. A. and BRONKHORST, A. W. (2000). "Fidelity of three-dimensional-sound reproduction using a virtual". In: *J. Acoust. Soc. Am.* 107.1, pp. 528–537 (cit. on p. 47).

LANGENDIJK, E. H. A. and BRONKHORST, A. W. (2002). "Contribution of spectral cues to human sound localization". In: *J. Acoust. Soc. Am.* 112.4, p. 1583 (cit. on p. 137).

LARCHER, V.; JOT, J.-M., and VANDERNOOT, G. (1998). "Equalization methods in binaural technology". In: *105th AES Conference*. Vol. 4858. San Francisco, USA (cit. on pp. 75, 121).

LENTZ, T. (2006). "Dynamic crosstalk cancellation for binaural synthesis in virtual reality environments". In: *J. Audio Eng. Soc.* 54.4, pp. 283–294 (cit. on pp. 3, 78, 93, 116).

LENTZ, T. (2007). "Binaural technology for virtual reality". PhD thesis. Institut für Technische Akustik, RWTH-Aachen (cit. on pp. 1, 3, 18, 21, 23, 47, 50, 51, 54–57, 59, 78, 94).

LI, Z. and DURAISWAMI, R. (2006). "Headphone-based reproduction of 3D auditory scenes captured by spherical/hemispherical microphone arrays". In: *IEEE International Conference on Acoustics, Speech, and Signal Processing*. Toulouse, France, pp. 337–340 (cit. on p. 2).

MACMILLAN, N. A. and CREELMAN, C. D. (2005). *Detection Theory*. Mahwah, New Jersey: LAWRENCE ERLBAUM ASSOCIATES, PUBLISHERS (cit. on pp. 108, 112).

MACPHERSON, E. A. and MIDDLEBROOKS, J. C. (2002). "Listener weighting of cues for lateral angle: The duplex theory of sound localization revisited". In: *J. Acoust. Soc. Am.* 111.5, p. 2219 (cit. on pp. 18, 137).

MAJDAK, P.; BALAZS, P., and LABACK, B. (2007). "Multiple exponential sweep method for fast measurement of head-related transfer functions". In: *J. Audio Eng. Soc.* 55.7/8, p. 623 (cit. on pp. 2, 21, 23, 33–37, 41, 63).

MAJDAK, P.; GOUPELL, M., and LABACK, B. (Feb. 2010). "3-D localization of virtual sound sources: effects of visual environment, pointing method, and training". In: *Atten. Percept. Psychophys.* 72.2, pp. 454–469 (cit. on pp. 14, 121).

MAJDAK, P.; GOUPELL, M., and LABACK, B. (2011). "Two-dimensional localization of virtual sound sources in cochlear-implant listeners". In: *Ear & Hearing* 32, pp. 198–208 (cit. on p. 126).

MAJDAK, P.; MASIERO, B., and FELS, J. (2012). "Human sound localization performance in individualized and non-individualized crosstalk cancellation systems". In: *J. Acoust. Soc. Am.* Submitted (cit. on p. 100).

MAMMONE, R. (Nov. 1999). "Inverse Problems and Signal Reconstruction". In: *Digital Signal Processing Handbook.* Ed. by V. K. Madisetti and D. B. Williams. Electrical Engineering Handbook. Boca Raton: CRC Press, pp. 1–4 (cit. on p. 7).

MASIERO, B. and FELS, J. (Mar. 2011a). "Equalization for Binaural Synthesis with Headphone". In: *Fortschritte der Akustik – DAGA.* Düsseldorf, Germany, pp. 675–676 (cit. on p. 67).

MASIERO, B. and FELS, J. (May 2011b). "Perceptually Robust Headphone Equalization for Binaural Reproduction". In: *130th AES Convention.* London, England, pp. 1–7 (cit. on pp. 67, 71).

MASIERO, B. and POLLOW, M. (May 2010). "A review of the compressive sampling framework in the lights of spherical harmonics: applications to distributed spherical arrays". In: *Second International Symposium on Ambisonics and Spherical Acoustics.* Paris, France (cit. on pp. 66, 149).

MASIERO, B. and QIU, X. (Mar. 2009). "Two Listeners Crosstalk Cancellation System Modelled by Four Point Sources and Two Rigid Spheres". In: *Acta Acustica united with Acustica* 95.2, pp. 379–385 (cit. on p. 82).

MASIERO, B. and VORLÄNDER, M. (2012). "A Framework for the Calculation of Dynamic Crosstalk Cancellation Filters". In: *IEEE Transactions on Audio, Speech and Language Processing.* Submitted (cit. on p. 77).

MASIERO, B.; POLLOW, M., and FELS, J. (June 2011a). "Design of a Fast Broadband Individual Head-Related Transfer Function Measurement System". In: *Proceeding of Forum Acusticum.* Aalborg, Denmark (cit. on pp. 21, 28).

MASIERO, B.; FELS, J., and VORLÄNDER, M. (2011b). "Review of the crosstalk cancellation filter technique". In: *Proceedings of the International Conference on Spatial Audio*. Ed. by M. Kob. Detmold, Germany (cit. on p. 77).

MASIERO, B.; POLLOW, M.; DIETRICH, P., and FELS, J. (2012). "Design of a Fast Measurement System for the Range Extrapolation of HRTFs". In: *J. Audio Eng. Soc.* Accepted (cit. on p. 21).

MIDDLEBROOKS, J. C. (Sept. 1999a). "Individual differences in external-ear transfer functions reduced by scaling in frequency." In: *J. Acoust. Soc. Am.* 106.3 Pt 1, pp. 1480–92 (cit. on p. 47).

MIDDLEBROOKS, J. C. (Sept. 1999b). "Virtual localization improved by scaling nonindividualized external-ear transfer functions in frequency." In: *J. Acoust. Soc. Am.* 106.3 Pt 1, pp. 1493–510 (cit. on pp. 14, 17, 19, 20, 126, 137).

MILLS, A. W. (1972). *Auditory localization*. Ed. by T. J. Vol. 2. New York Academic Press (cit. on p. 19).

MINNAAR, P.; FLEMMING, C.; MØLLER, H.; OLESEN, S. K., and PLOGSTIES, J. (1999). "Audibility of All-Pass Components in Binaural Synthesis". In: *106th AES Convention* (cit. on p. 73).

MOKHTARI, P.; TAKEMOTO, H.; NISHIMURA, R., and KATO, H. (2007). "Comparison of Simulated and Measured HRTFs: FDTD Simulation Using MRI Head Data". In: *123rd AES Convention* (cit. on p. 2).

MØLLER, H. (1992). "Fundamentals of binaural technology". In: *Applied Acoustics* 36.3-4, pp. 171–218 (cit. on pp. 1, 3, 67–69).

MØLLER, H.; SØRENSEN, M. F.; HAMMERSHØI, D., and JENSEN, C. B. (1995a). "Head-Related Transfer Functions of Human Subjects". In: *J. Audio Eng. Soc.* 43.5, pp. 300–310 (cit. on pp. 2, 13, 21, 23, 25).

MØLLER, H.; HAMMERSHØI, D.; JENSEN, C. B., and SØRENSEN, M. F. (1995b). "Transfer Characteristics of Headphones Measured on Human Ears". In: *J. Audio Eng. Soc.* 43.4, pp. 203–217 (cit. on pp. 3, 22, 68).

MØLLER, H.; SØRENSEN, M. F.; JENSEN, C. B., and HAMMERSHØI, D. (1996). "Binaural technique: Do we need individual recordings?" In: *J. Audio Eng. Soc.* 44.6, pp. 451–469 (cit. on pp. 13, 101).

MOORE, B. C. J. (2012). *An Introduction to the Psychology of Hearing*. Emerald Group (cit. on p. 104).

MORIMOTO, M. and AOKATA, H. (1984). "Localization cues in the upper hemisphere". In: *J Acoust Soc Jpn (E)* 5, pp. 165–173 (cit. on pp. 18, 119).

MÜLLER, S. (1999). "Digitale Signalverarbeitung für Lautsprecher". Ph.D. RWTH Aachen University, p. 263 (cit. on p. 72).

MÜLLER, S. (2008). "Measurement of Transfer Functions and Impulse Responses". en. In: *Handbook of Signal Processing in Acoustics*, ed. by D. Havelock; S. Kuwano, and M. Vorländer. Vol. -1. New York, NY: Springer New York, pp. 65–85 (cit. on p. 10).

MÜLLER, S. and MASSARANI, P. (2001). "Transfer-function measurement with sweeps". In: *J. Audio Eng. Soc.* 49.6, pp. 443–471 (cit. on pp. 10, 11, 33).

NELSON, P.; KIRKEBY, O.; TAKEUCHI, T., and HAMADA, H. (July 1997). "Sound Fields for the Production of Virtual Acoustic Images". In: *Journal of Sound and Vibration* 204.2, pp. 386–396 (cit. on p. 132).

NELSON, P. A. and ELLIOTT, S. J. (1995). *Active Control of Sound*. 3rd. San Diego, CA: Academic Press, p. 436 (cit. on pp. 8, 155, 156).

NELSON, P. A. and ROSE, J. F. W. (Sept. 2006). "The time domain response of some systems for sound reproduction". In: *Journal of Sound and Vibration* 296.3, pp. 461–493 (cit. on pp. 77, 89).

NORCROSS, S. G. and BOUCHARD, M. (2007). "Multichannel Inverse Filtering with Minimal-Phase Regularization". In: *123rd AES Convention*, pp. 1–8 (cit. on pp. 9, 90–92).

OBEREM, J. (2012). "Analysis of different equalization methods for binaural reproduction". M.Sc. RWTH Aachen University (cit. on pp. 69, 99).

OPPENHEIM, A. and SCHAFER, R. (1989). *Discrete-Time Signal Processing*. 1st. Vol. Signal Pro. Prentice Hall (cit. on pp. 5–7).

PAPOULIS, A. (1977). *Signal Analysis*. McGraw-Hill, p. 431 (cit. on p. 88).

PAQUIER, M. and KOEHL, V. (2010). "Audibility of headphone positioning variability". In: *128th AES Convention* (cit. on p. 69).

PARODI, Y. L. (2008). "Analysis of design parameters for crosstalk cancellation filters applied to different loudspeaker configurations". In: *125th AES Convention*. San Francisco, USA (cit. on p. 78).

PARODI, Y. L. (2010). "A Systematic Study of Binaural Reproduction Systems Through Loudspeakers: A Multiple Stereo-Dipole Approach". PhD. Aalborg University (cit. on p. 84).

PARODI, Y. L. and RUBAK, P. (2011). "A Subjective Evaluation of the Minimum Channel Separation for Reproducing Binaural Signals over Loudspeakers". In: *J. Audio Eng. Soc.* 59.7-8, pp. 487–497 (cit. on pp. 117, 134, 141).

PARODI, Y. L. and RUBAK, P. (Sept. 2010). "Objective evaluation of the sweet spot size in spatial sound reproduction using elevated loudspeakers." In: *J. Acoust. Soc. Am.* 128.3, pp. 1045–55 (cit. on pp. 83, 93).

PAUL, S. (Sept. 2009). "Binaural Recording Technology: A Historical Review and Possible Future Developments". In: *Acta Acustica united with Acustica* 95.5, pp. 767–788 (cit. on pp. 2, 10).

PELTONEN, T. (2000). "A Multichannel Measurement System for Room Acoustics Analysis". M.Sc. Helsinki University of Technology (cit. on p. 11).

PELZER, S.; MASIERO, B., and VORLÄNDER, M. (2011). "3D Reproduction of Room Acoustics using a Hybrid System of Combined Crosstalk Cancellation and Ambisonics Playback". In: *Proceedings of the International Conference on Spatial Audio.* Ed. by M. Kob. Detmold, Germany, pp. 297–301 (cit. on p. 150).

PERRETT, S and NOBLE, W (Oct. 1997). "The effect of head rotations on vertical plane sound localization." In: *J. Acoust. Soc. Am.* 102.4, pp. 2325–32 (cit. on p. 137).

POLLOW, M.; NGUYEN, K.-V.; WARUSFEL, O.; CARPENTIER, T.; MÜLLER-TRAPET, M.; VORLÄNDER, M., and NOISTERNIG, M. (Jan. 2012a). "Calculation of Head-Related Transfer Functions for Arbitrary Field Points Using Spherical Harmonics Decomposition". In: *Acta Acustica united with Acustica* 98.1, pp. 72–82 (cit. on pp. 24, 47, 50, 51, 63).

POLLOW, M.; MASIERO, B.; DIETRICH, P.; FELS, J., and VORLÄNDER, M. (2012b). "Fast Measurement System for Spatially Continuous Individual HRTFs". In: *Spatial Audio in Today's 3D World - AES 25th UK Conference*, pp. 1–8 (cit. on p. 21).

POLLOW, M.; DIETRICH, P.; MASIERO, B.; FELS, J., and VORLÄNDER, M. (2012c). "Modal sound field representation of HRTFs". In: *Fortschritte der Akustik – DAGA.* Darmstadt, Germany (cit. on p. 65).

PRALONG, D and CARLILE, S. (Dec. 1996). "The role of individualized headphone calibration for the generation of high fidelity virtual auditory space." In: *J. Acoust. Soc. Am.* 100.6, pp. 3785–93 (cit. on p. 3).

PULKKI, V. (1997). "Virtual sound source positioning using vector base amplitude panning". In: *J. Audio Eng. Soc.* 45.6, pp. 456–466 (cit. on pp. 1, 141).

QIU, X.; MASIERO, B., and VORLÄNDER, M. (2009). "Experimental Study on Channel Separation of Crosstalk Cancellation System with Mismatched Sound Sources". In: *The 10th Western Pacific Acoustics Conference.* Beijing, China (cit. on pp. 83, 117).

QU, T.; XIAO, Z.; GONG, M.; HUANG, Y.; LI, X., and WU, X. (Aug. 2009). "Distance-Dependent Head-Related Transfer Functions Measured With High Spatial Resolution Using a Spark Gap". In: *IEEE Transactions on Audio, Speech and Language Processing* 17.6, pp. 1124–1132 (cit. on p. 24).

RAO, H. I. K.; MATHEWS, V. J., and PARK, Y.-C. (Nov. 2007). "A Minimax Approach for the Joint Design of Acoustic Crosstalk Cancellation Filters". In: *IEEE Transactions on Audio, Speech and Language Processing* 15.8, pp. 2287–2298 (cit. on p. 86).

RAYLEIGH, L. (Feb. 1907). "XII. On our perception of sound direction". In: *Philosophical Magazine Series 6* 13.74, pp. 214–232 (cit. on p. 18).

RUFFINI, G.; MARCO, J., and GRAU, C. (2002). *Spherical harmonics interpolation, computation of Laplacians and Gauge Theory.* arXiv: 0206007v1 [arXiv:physics] (cit. on pp. 50, 157).

RUI, Y.; YU, G., and XIE, B. (Sept. 2012). "Approximately calculate individual near-field head-related transfer function using an ellipsoidal head and pinnae model." en. In: *J. Acoust. Soc. Am.* Vol. 132. 3, p. 1997 (cit. on p. 2).

RUMSEY, F. (2012). *Spatial Audio.* CRC Press, p. 256 (cit. on p. 1).

SÆBØ, A. (2001). "Influence of Reflections on Crosstalk Cancelled Playback of Binaural Sound". Ph.D. Norwegian University of Science and Technology (cit. on p. 97).

SARTOR, M. (2010). "Entwurf von Lautsprecher und Messbogen zur HRTF-Messung". Studienarbeit. RWTH Aachen (cit. on p. 26).

SCHMIDT, S. (2009). "Finite Element Simulation of External Ear Sound Fields for the Optimization of Eardrum-Related Measurements". Ph.D. Ruhr-Universität Bochum, p. 145 (cit. on pp. 67, 71).

SCHONSTEIN, D.; FERR, L., and KATZ, B. F. G. (2008). "Comparison of headphones and equalization for virtual auditory source localization". In: *Acoustics '08, Paris* (cit. on p. 121).

SCHRÖDER, D.; WEFERS, F.; PELZER, S.; RAUSCH, D. S.; VORLÄNDER, M., and KUHLEN, T. (2010). "Virtual Reality System at RWTH Aachen University". In: *Proceedings ICA 2010, 20th International Congress on Acoustics :* ed. by M. Burgess. Sydney, Australia: Australian Acoustical Society, NSW Division (cit. on p. 1).

SCHROEDER, M. R. (Aug. 1979). "Integrated-impulse method measuring sound decay without using impulses". en. In: *J. Acoust. Soc. Am.* 66.2, p. 497 (cit. on p. 11).

SEEBER, B. (2002). "Untersuchung der auditiven Lokalisation mit einer Lichtzeigermethode". Ph.D. TU München (cit. on p. 18).

SLEPIAN, D (1964). "Prolate spheroidal wave functions, Fourier analysis and uncertainty–IV: Extension to Many Dimensions; Generalized Prolate Spherical Functions". In: *Bell Syst. Tech. J* 43.6, pp. 3009–3057 (cit. on p. 66).

TAKEUCHI, T. and NELSON, P. A. (2007). "Subjective and objective evaluation of the optimal source distribution for virtual acoustic imaging". In: *J. Audio Eng. Soc.* 55.11, p. 981 (cit. on pp. 3, 77, 86, 95, 117).

TAKEUCHI, T.; NELSON, P. a., and HAMADA, H. (2001). "Robustness to head misalignment of virtual sound imaging systems". In: *J. Acoust. Soc. Am.* 109.3, p. 958 (cit. on pp. 116, 117).

TOHYAMA, M. and KOIKE, T. (1998). *Fundamentals of Acoustic Signal Processing.* Academic Press (cit. on pp. 5, 13).

TOOLE, F. E. (1984). "The Acoustics and Psychoacoustics of Headphones". In: *2nd AES International Conference.* Anaheim, USA (cit. on p. 69).

VANDERKOOY, J. (Nov. 2010). "Rapid In-Place Measurements of Multichannel Venues". In: *129th AES Convention* (cit. on p. 33).

VÖLK, F. (2011a). "Inter- and Intra-Individual Variability in blocked auditory canal transfer functions of three circum-aural headphones". In: *131st AES Convention* 8465, p. 10 (cit. on p. 69).

VÖLK, F. (2011b). "System Theory of Binaural Synthesis". In: *131st AES Convention* 8568, p. 17 (cit. on p. 69).

DE VRIES, D. (1988). *Wave Field Synthesis*. Audio Engineering Society (cit. on p. 1).

WANG, L.; YIN, F., and CHEN, Z. (2009). "Head-related transfer function interpolation through multivariate polynomial fitting of principal component weights". In: *Acoustical Science and Technology* 30.6, pp. 395–403 (cit. on p. 47).

WEINZIERL, S.; GIESE, A., and LINDAU, A. (2009). "Generalized Multiple Sweep Measurement". In: *126th AES Convention* (cit. on p. 36).

WENZEL, E. M.; ARRUDA, M.; KISTLER, D. J., and WIGHTMAN, F. L. (July 1993). "Localization using nonindividualized head-related transfer functions". In: *J. Acoust. Soc. Am.* 94.1, pp. 111–123 (cit. on pp. 2, 13, 14).

WIGHTMAN, F. L. and KISTLER, D. J. (Feb. 1989). "Headphone simulation of free-field listening. II: Psychophysical validation". In: *J. Acoust. Soc. Am.* 85.2, pp. 868–878 (cit. on pp. 2, 13, 101).

WIGHTMAN, F. L. and KISTLER, D. J. (1992). "The dominant role of low-frequency interaural time differences in sound localization." In: *J. Acoust. Soc. Am.* 91.3, pp. 1648–1661 (cit. on p. 137).

WILLIAMS, E. G. (1999). *Fourier Acoustics: sound radiation and nearfield acoustical holography*. Academic Press (cit. on pp. 49, 51).

XIANG, N and SCHROEDER, M. R. (2003). "Reciprocal maximum-length sequence pairs for acoustical dual source measurements". In: *J. Acoust. Soc. Am.* 113, p. 2754 (cit. on p. 33).

XIE, B. and ZHANG, T (2010). "The Audibility of Spectral Detail of Head-Related Transfer Functions at High Frequency". In: *Acta Acustica united with Acustica* (cit. on pp. 47, 149).

YANG, J.; GAN, W., and TAN, S.-E. (2003). "Improved sound separation using three loudspeakers". In: *Acoustic Research Letters* 4.April, pp. 47–52 (cit. on p. 93).

ZHANG, W.; ZHANG, M., and KENNEDY, R. (2012). "On High-Resolution Head-Related Transfer Function Measurements: An Efficient Sampling Scheme". In: *IEEE Transactions on Audio, Speech and Language Processing* 20.2, pp. 575–584 (cit. on pp. 30, 31, 43, 49, 65).

ZIEGELWANGER, H. (2012). "Efficient modeling of the time-of-arrival in binaural reproduction of virtual sound sources". Diplom. Universit für Musik und darstellende Kunst Graz, p. 98 (cit. on pp. 29, 45, 149).

ZOTKIN, D. N.; DURAISWAMI, R.; GRASSI, E., and GUMEROV, N. A. (2006). "Fast head-related transfer function measurement via reciprocity". In: *J. Acoust. Soc. Am.* 120.4, p. 2202 (cit. on pp. 2, 21, 22).

ZOTTER, F. (Jan. 2009). "Analysis and Synthesis of Sound-Radiation with Spherical Arrays". Ph.D. University of Music and Performing Arts, Graz, Austria (cit. on pp. 24, 31, 49, 65).

ZOTTER, F. (2010). "Sampling Strategies for Acoustic Holography/Holophony on the Sphere". In: *Fortschritte der Akustik – DAGA* (cit. on pp. 31, 50).

Acknowledgments

It is now time to thank and acknowledge the many helping hands that led to the conclusion of this thesis.

First, I'm obliged to thank the Brazilian National Council for Scientific and Technological Development (CNPq) for the financial support and for making it possible for me to conduct my research in a highly renowned institution in the field of acoustics.

I would like to thank Prof. Dr. rer. nat. Michael Vorländer, head of the Institute of Technical Acoustics (ITA) of the RWTH Aachen University, for welcoming me to his institute and accepting to take over the role of my scientific adviser. Your ability to disseminate your wisdom and your determination to expand the field of acoustics have impressed me very much.

I also would like to thanks Prof. Philip Nelson, FREng. for accepting the invitation to be the second referee for this thesis and providing insightful comments on the thesis.

Thanks to Dr. Gottfried Behler for his willingness to discuss any problem that might occur and almost always coming up with a practical solution. And thanks to Prof. Dr. Ing. Janina Fels for giving me the opportunity to stay one year longer at ITA participating in one of her projects and for the many suggestions and advice in the final stages of this dissertation.

During the first year of this thesis I was lucky to work with Dr. Xiaojun Qiu. Thank you for the very enriching collaboration. And during the last year of my thesis I was again lucky to work with Dr. DI Pjotr Majdak at the Acoustics Research Institute of the Austrian Academy of Sciences. I would like to express my greatest gratitude for you having me there and for the great collaboration. Also many thanks to Michael Mihocic for running the localization tests and helping out in all possible ways.

Special thanks go to the staff of the mechanical and electronic workshops at ITA, namely Rolf Kaldenbach, Hans-Jürgen Dilly, Uwe Schlömer, Thomas Schäfer, and all apprentices, for the outstanding work of bringing to life ideas that otherwise would have stayed only on paper; and to the institute's secretariat, namely Ulrike Görgens (*in memoriam*),

Wilma Vonhoegen, and Karin Charlier, for making our life easier and less bureaucratic.

It is said that "a chain is only as strong as its weakest link". I'm glad I was in an environment with colleagues that were very "strong" in the very many different field of acoustics. I am greatly indebted to some colleagues who made a substantial contribution to this thesis: to Martin Pollow for endless discussions about all aspects of dealing with "spherical acoustics" and for the words of advice about doing a "round the world" trip; to Pascal Dietrich who helped in all matters related to acoustic measurement and initiated all self-help groups we had at ITA; and to the VR-Crew Frank Wefers and Sönke Pelzer who always supported me in the many challenges of spatial audio reproduction. Special thanks to Johannes Klein for helping with the daunting dark pictures. For all other colleagues—Dr. Andreas Frank, Dr. Dirk Schröder, Elena Shabalina, Dr. Elzbieta Nowicka, Ingo Witew, Jan Köhler, Dr. Marc Aretz, Markus Müller-Trappet, Martin Guski, Matthias Lievens, Oliver Strauch, Ramona Bomhardt, Renzo Vitale, Rob Opdam, Roman Scharrer, Dr. Sebastian Fingerhuth, and Xun Wang—thanks for the friendly and productive atmosphere.

I also would like to thank Dr. Marcio de Avelar Gomes for opening up the first opportunity for me to come to Aachen and Dr. João Henrique Diniz Guimarães for all the help in my first days in Aachen.

During the five year I spent in ITA I had the chance to work with many very skilled students who contributed in one way of another to the development of this thesis. Björn Kutzner, Souptik Barua, Srikanth Korse, Thomas Bierbaums, Gengliang Han, Malte Sartor, Jörg Seidler, Mario Otten, Jochen Giese, Johnny Nahas, Johannes Fundalewicz, Benedikt Koppers Benedikt Krechel. I would like to thank every single one of you for your trust and hard work. Special thanks to Josefa Oberem for so orderly conducting the perceptual tests.

On a more personal level, I would like to thank all players of the Rugby Club Aachen for providing me with such a good atmosphere where I could (almost always) forget about the pressures of my daily work.

Being away from home is not always easy. The Pfaff family was my step-family throughout these years. They were always there when I needed them, be it when I broke a tooth or had a shoulder surgery or just when I did not have any plans for Easter.

Last, but above all, I would like to thank my family. Even though you were not always physically present, you were always there for me. Special thanks goes to my father for always asking when I was going to be done with this dissertation, for proof-reading and greatly improving the only parts of this dissertation that will probably be read by anyone else, and for introducing me to the world of buying wine. I would also like to thank my little sister for talking with me from the other side of the world when I was just feeling down and for making sure that I did not work all through the night just before the deadline. And to my mother for being the loving person that holds our family that is spread all over the world together. This thesis is dedicated to the three of you!

Curriculum Vitæ

Personal Data

	Bruno Sanches Masiero
20.12.1981	born in São Paulo, Brazil
	as son of Vera Lúcia und Paulo Masiero

Education

02/1989–12/1996	Primary School in São Carlos (SP), Brazil
02/1997–12/1999	Secondary School in São Carlos (SP), Brazil

Higher Education

02/2000–12/2005	Bachelor's degree in Electrical Engineering at Universidade de São Paulo, Brazil Major: Telecommunication
10/2004–06/2005	Free mover at the RWTH Aachen University
01/2006–07/2007	Master's degree in Electrical Engineering at Universidade de São Paulo, Brazil Specialization: Electronic Systems

Employments

06/2003–05/2004	Student worker at the Department of Music, Universidade de São Paulo, Brazil
06/2004–09/2004	Internship at Siemens AG, Munich
02/2005–05/2005	Student worker at the Institute of Technical Acoustics, RWTH Aachen University
10/2007–09/2011	Stipendiary from CNPq, Brazil, at the Institute of Technical Acoustics, RWTH Aachen University
10/2011–10/2012	Research Assistant at the Institute of Technical Acoustics, RWTH Aachen University

Bisher erschienene Bände der Reihe

Aachener Beiträge zur Technischen Akustik

ISSN 1866-3052

Alle erschienenen Bücher können unter der angegebenen ISBN-Nummer direkt online
(http://www.logos-verlag.de) oder per Fax (030 - 42 85 10 92) beim Logos Verlag
Berlin bestellt werden.